Contributions to Management Science

The series *Contributions to Management Science* contains research publications in all fields of business and management science. These publications are primarily monographs and multiple author works containing new research results, and also feature selected conference-based publications are also considered. The focus of the series lies in presenting the development of latest theoretical and empirical research across different viewpoints.

This book series is indexed in Scopus.

Gianfranco Minati · Maria Pietronilla Penna
Editors

Multiple Systems

Complexity and Coherence in Ecosystems, Collective Behavior, and Social Systems

Springer

Editors
Gianfranco Minati
AIRS/Italian Systems Society
Milano, Italy

Maria Pietronilla Penna
University of Cagliari
Cagliari, Italy

ISSN 1431-1941 ISSN 2197-716X (electronic)
Contributions to Management Science
ISBN 978-3-031-44684-9 ISBN 978-3-031-44685-6 (eBook)
https://doi.org/10.1007/978-3-031-44685-6

This Springer imprint is published by the registered company Springer Nature Switzerland AG
The registered company address is: Gewerbestrasse 11, 6330 Cham, Switzerland

In memory of Professor Eliano Pessa

September 19, 1946–March 22, 2020

Eliano Pessa, Theoretical Physicist, was a Full Professor of General Psychology and Cognitive Modeling at the University of Pavia, Italy. He has already been Dean of the Department of Psychology and the Inter-departmental Research Center on Cognitive Science in the same university. He has also previously held Associate Professor of Artificial Intelligence at the University of Rome "La Sapienza" Faculty of Psychology. He was the author or coauthor of 10 books and many papers in scientific journals, books, and proceedings of international conferences. His scientific research interests included quantum theories describing the human brain's operation, computational neuroscience, artificial neural networks, modeling emergence processes, quantum field theory, phase transitions in condensed matter, human memory, visual perception, decision-making, and statistical reasoning. As an expert mountaineer, he has climbed mountains on various continents.

Scientific Committee

Preface

The Eighth National Conference of the Italian Systems Society is held in conjunction with its 27th year of activity after its foundation.

The congress is held after an unusual period of closure due to the COVID problem and the loss of Professor Eliano Pessa, a loss that marked everything and all of us for its emotional and existential impact.

This congress is dedicated to the memory of Professor Eliano Pessa, our dear friend, vice-president of the Italian Systems Society, and greatest systems scientist. His gentleness and modesty were second only to his very high cultural and scientific levels.

It's our turn to face the shame of survival. We will try not to be only devourers of moments, but to be aware of time and honor it.

The title of the conference, *Multiple Systems*, aims to underline the need for systemics and systems science to deal with equivalent and non-equivalent multiplicity of systems and quasi-systems of complexity. The theme of multiple systems, along with many others, had been elaborated and discussed with Eliano, shortly before his disappearance. The choice of the theme is in honor of him, all of us grateful for his innumerable contributions whose outstanding level was second only to his modesty and meekness.

The subject is interdisciplinary elaborated by various contributions and is looking forward for further research.

The topic of this eight conference is an evolution of the subjects of previous conferences, all edited with Eliano, namely

- 2002 *Emergence in Complex Cognitive, Social and Biological Systems.*
- 2006 *Systemics of Emergence: Research and Applications.*
- 2009 *Processes of Emergence of Systems and Systemic Properties-Towards a General Theory of Emergence.*
- 2012 *Methods, Models, Simulations and Approaches Towards a General Theory of Change.*
- 2016 *Towards a Post-Bertalanffy Systemics.*
- 2019 *Systemics of Incompleteness and Quasi-Systems.*

The topic of multiple systems is considered in the literature from different points of view such as multiple systems of care, multiple system atrophy or multisystem atrophy neurodegenerative pathology, multiple levels of processing in decision making, polypathologies, and others. The theme does not concern systems made up of subsystems, in a conceptual decomposition that is often only linear and functional.

The issue we consider here concerns the occurrence of multiple interactions related to (a) the ability of a generic agent to both interact with other agents using dynamic and context-sensitive combinations of specific interaction rules, and (b) contextual multiple roles or multiple meanings of results produced by specific interactions. The issue arises for complex systems, collective behaviors, and networks in which there are interchangeability and dynamic equivalences, however with the consistency of multiple coherences, as in (Minati 2021; 2022a).

However, this simplified situation must also be extended considering that:

1. Agents can belong to different systems at the same time, for example by managing different interactions simultaneously as nodes in networks with different connections, or when the same interactions have different effects when combined with others.
2. Agents can dynamically give rise to subsequent different systems.

Furthermore, the concept of interaction must actually be specified (Minati 2022b) at least as regards some aspects such as:

1. The interaction consists of the exchange of matter and energy processed in a standard way (reactions as in the case of collisions between balls) and information that the elements process in variable (e.g., adaptive) modes. Interacting in populations of elements can consist of multiple and overlapping cases.
2. The interacting elements instead of simplistic fixed pairs can be variable pairs, consisting of clusters and systems that process the interaction in a variable way.
3. The occurrence in a way that is not always constant and identical, incomplete and according to any regularity as in the case of quasi-systems (Minati and Pessa 2018).
4. The occurrence in combinations, in their turn variable of combinations of interactions.
5. The interaction between interactions as in the case of interference.
6. Furthermore, the distinction between elements and interactions and the need to present themselves with specific standardized roles is simplistic when the roles are not fixed and well defined.

Furthermore, the following must be considered:

1. The interactants or inter-actors who generate and process the interactions such as agents, clusters, and systems.
2. Interactions of a duration and life less than that of inter-actors.
3. The presence of interaction domains, the occurrence of domains as fields of possible interactions, and their mutual influences such as those belonging to the basin of an attractor. The entry of an entity into these domains, implicitly full of

specific possibilities, significantly influences its behavior, as does entering into a collective behavior (Minati 2019).
4. Remote interactions (such as remote synchronization phenomena) which occur, for example, when pairs of non-adjacent entities become substantially synchronized despite the absence of direct structural connections between them or intermediate mediating entities such as in the brain and networks (Gambuzza *et al.* 2013; Minati 2015).
5. Aspects of incompleteness of interactions.
6. Role of dissipation.
7. Role of ergodicity when the same system can be both ergodic and non-ergodic depending on the timescale of the observer, as in polymers, or even temporarily ergodic (Minati and Pessa 2018, p. 69, p. 169).

It is a scenario typically identifiable in ecosystems, systems, and social networks (e.g., behaving as members of families, buyers, workers, members of the traffic system, users of services, listeners of programs), complex systems, and collective behaviors. For example, in network systems where different nodes are sharing the same links (as in the Internet); in living systems where the same interactions have different roles and meanings such as biochemical and psychological.

It is a question of identifying the role of multiple systems in complexity, emergence, and the establishment of coherences as in collective systems.

The creation of multiple systems is of particular interest as it allows to influence of one or more systems that are difficult to manage by acting on another or others that are easier to influence and access.

This is necessary to act on systems with high structural dynamics by varying, for example, the way they interact, subsequent structural changes such as for the cytoskeleton and for complex systems that can be understood as sequences of phase transitions where the properties of such sequences should be understood as a structural dynamic, coherent in complex systems (Minati and Licata 2013). Several possible cases can occur separately or together and in any combination:

- Change of structure, i.e., from one structure to another.
- Acquisition of a structure, or passage from an unstructured to a structured configuration.
- Loss of structure, i.e., transition from a structured to an unstructured configuration.
- Combinations of structures.

The multiplicity of Multiple Systems cannot be reduced to multiplicity of different aspects, as mere fact of relativism observer-dependent. The system is, rather, multiple as established by multiple levels of coherence and emergence.

This opens the door to the interdisciplinary understanding of the concept when considering, for instance, the *integrated multiplicity* of systems such as architectural, cultural, economic, linguistic, medical, philosophical, psychological, religious, scientific, and social systems. An alleged General Systems Theory ignoring multiple systems should be intended as a case of reductionism having theoretical and practical effects.

As examples we mention the incredible ineffectiveness and danger (e.g., uncontrolled manipulations) of not considering multiple systems in the field of social sciences, for example planning and undertaking political initiatives, in the inability to resolve conflicts; of economic sciences and in the field of management ignoring multiple and overlapping effects in managing corporate and social systems.

This conference aims to explore cases and present conceptual approaches within the conceptual context described above as a thematic of theoretical systems, e.g., in architecture, biology, economics, management, mathematics, medicine, philosophy, physics, psychology, and sociology.

The contributions explored cases and presented conceptual approaches within the novel context described above.

We conclude by observing how this approach is consequential to issues developed in previous congresses such as those of incompleteness and quasi-systems (Minati *et al.* 2019).

Milano, Italy — Gianfranco Minati, AIRS president

Cagliari, Italy — Maria Pietronilla Penna, Co-Editor
July 2023

References

1. Minati, G. (Ed.). (2021). Multiplicity and interdisciplinarity—essays in honor of Eliano Pessa. Springer.
2. Minati, G. (2022a). Multiplicity, logical openness, incompleteness, and quasi-ness as peculiar non-reductionist properties of complexity. In S. Wuppuluri, I. Stewart (Eds.) *From Electrons to Elephants and Elections: Saga of Content and Context* (pp. 151–173). Springer.
3. Minati, G. (2022b). Interactions and interaction mechanisms as generators of complex systems. *WSEAS Transactions on Proof*, *2*, 96–107. https://wseas.com/journals/articles.php?id=5545
4. Minati, G., Pessa, E. (2018). From collective beings to quasi-systems. Springer.
5. Minati, G. (2019). Non-classical systemics of quasi-coherence: from formal properties to representations of generative mechanisms. A conceptual introduction to a paradigm-shift. *Systems*, *7*(4). https://doi.org/10.3390/systems7040051
6. Gambuzza, L. V., Cardillo, A., Fiasconaro, A., Fortuna, L., Gómez-Gardenes, J., Frasca, M. (2013). Analysis of remote synchronization in complex networks. *Chaos*, *23*, 1–8.
7. Minati, L. (2015). Remote synchronization of amplitudes across an experimental ring of non-linear oscillators. *Chaos*, *25*, 123107–123112.
8. Minati, G., Licata, I. (2013). Emergence as mesoscopic coherence. *Systems*, *1*(4), 50–65.https://doi.org/10.3390/systems1040050
9. Minati, G., Abram, M., Pessa, G. (Eds). (2019). Systemics of incompleteness and quasi-systems. Springer.

Acknowledgements

The Eighth Italian Conference has been possible thanks to the contributions of many people who have accompanied and supported the growth and development of AIRS during all the years since its establishment and to the contribution of "new" energies from students and researchers realizing the systemic aspects of their activity.

We thank the University of Cagliari, the President of the Faculty of Humanities Prof. Antonello Mura, the Psychology Program Coordinator (bachelor's degree and master's degree in psychology) Prof. Cristina Cabras for welcoming and hosting host this event.

We have been honored by the presence of Prof. Giovanni Biggio, emeritus professor of Neuropsychopharmacology, University of Cagliari, who delivered the opening plenary lecture "Epigenetics and brain development: how the environment changes the brain over the course of life."

We have been honored by the opening plenary lecture "Neural networks and many-body systems" delivered by Prof. Giuseppe Vitiello, emeritus professor, Department of Physics "E. R. Caianiello," University of Salerno.

We thank all the authors who submitted papers for this conference and in particular the members of the program committee as well as the referees who have guaranteed the quality of the event.

We thank explicitly all the people who have contributed during the conference, bringing ideas and stimuli to this new phase of the scientific and cultural project of systemics.

Contents

About the Editors

Gianfranco Minati systems scientist, has switched from a position as executive in a large industrial-financial Italian group (1979–1984) to research. He is founder (1996) and president of the Italian Systems Society (AIRS); member of the scientific committee of Conferences and Systems Societies; doctoral lecturer (2000–2017) at the Polytechnic of Milan, Italy. Springer author, he is author of 39 chapters in books; editor of 11 books; author or co-author of 16 books; author of 59 articles and academic publications. His current research interest focuses on artificial unconscious in AI, complex systems, Dynamic Usage of Models (DYSAM), emergence, logical openness, mesoscopic coherence, meta-structures, multiple systems, quasi-systems, theoretical incompleteness, architecture and design as social meta-structures to influence processes of emergence in social systems.

Maria Pietronilla Penna full professor in General Psychology at the University of Cagliari. She is member Italian Systems Society (AIRS) since the very beginning. Her current research interest focuses on artificial intelligence, neural network models, systems theory, cognitive psychology, psychophysiology, psychology of communication, man-machine interfaces, usability assessment, e-learning design, memory, perception and mental representations, connectionist models of cognitive processing, iconic memory, pattern perception, mostly in the case of ambiguous pictures, perceptive field structuring in homogeneous stimulation areas, spatial memory, movement perception, problem solving and learning processes in school-age children, e-learning tools usability and design, text comprehension in school children, communication processes, aging, addictions, technology use. She worked on perception, memory, knowledge representation, both through laboratory experiments and theoretical formulation of new cognitive process models. She's author of author of numerous publications.

Multiplicity

Multiple Systems

Gianfranco Minati

Abstract The term "multiple systems" was coined to describe the occurrence of multiple, variable interactions involving commonly the same elements and their interchangeability and equivalence, as seen in ecosystems, collective behaviors, and multiple networks. It also relates to the complex dynamics of emergence, such as multiple, partial, tentative, and failing, for instance, in social systems where sequences of acquisitions of multiple roles and properties like cultural, consumerist, familial, political, professional, and religious for single agents overlap and combine one another. More broadly, multiple systems are related to structural dynamics, where the dynamics are related to changes in interaction structures (the way interactions are arranged, organized, or interfere) rather than changes in values regarding the same interaction structure, and different possible cases can occur separately or simultaneously in any combination. In contrast with "complete-able incompleteness", the changing of multiple interaction structures is intended to be characterized by "theoretical incompleteness," a property suitable to leave the system provided by such a property in a state of permanent indefiniteness given by multiplicities of combined elements and interactions with parametrical and timing values that are predominantly always different. The multiple versions of such systems are conceivable as a network of possibilities, from which configurations endowed with partial and dynamic coherence are selected, for instance, due to environmental disturbances and the occurrence of random prevalence. Finally, the presence of potential "superimposed" multiple systems associated with pending, inactivated, implicit interactions waiting for suitable environmental conditions, e.g., energetic such as for metastability, to "collapse" in the activation of a specific version, can be considered. This is the case in ecosystems and social systems. Interestingly, the contrast between classical and quantum collapse can be considered. A General Systems Theory ignoring multiple systems should be intended as a case of reductionism having theoretical and practical effects.

G. Minati (✉)
Italian Systems Society Via Pellegrino Rossi, 42B 20161 Milano, Italy
e-mail: gianfranco.minati@AIRS.it

G. Minati and M. Pietronilla Penna (eds.), *Multiple Systems*, Contributions to Management Science, https://doi.org/10.1007/978-3-031-44685-6_1

1 Introduction

The concept of multiplicity (Minati, 2021a) relates to phenomena and modeling where despite the dynamics of maintaining, losing, and restoring coherence levels, typical of complexity, there are multiple predominant equivalences, interchangeability, superimpositions, and coexisting options to be *decided* or *resolved*, for instance, through environmental perturbations and internal random fluctuations.

In Sect. 2, we consider multiple systems as defined by the variable multiplicities of interacting elements, ranging from not being regularly active, much less identical in reacting, to being gradually replaced by partially equivalent elements, until partial interchangeability allows configurations of admissible compatible dynamics.

When considering the variable multiplicities of interactions, the situation repeats itself, e.g., in terms of their occurring in variable arrangements; in interesting not only well-defined, fixed groups of elements (the interactive and the interactants), but multiple variable momentary groups; in occurring at different starting times and having variable duration; in changing in properties such as intensity; and in being the result of previous combinations (Minati, 2022a).

We focus on the concept of structural dynamics. A structure is usually intended, in general, as the way in which the parts of a system are arranged, organized, or regularly interact. In addition, the dynamics of structural dynamics relate to the changing of the arrangement, organizations, and interference between interactions, such as the variable interaction mechanism of the cytoskeleton (Minati & Pessa, 2018, pp. 87–90), rather than the changing of values regarding the *same* interaction structure.

The general systemic identity is compatible, tolerant of reciprocally suitable levels of structural variations such as those in communities, e.g., families, working places, and schools; assemblages using compatible composing entities; living systems that restructure themselves cellularly, e.g., in the process of self-repairing; shape and topological changes in collective systems, e.g., swarms and flocks; and self-organized systems, e.g., whirlpools. More than compatible, the modalities by which such structural variations occur provide the systemic identity.

In Sect. 3, we consider cases in which this structural dynamic of elements and interactions allows the incompleteness (Minati, 2016; Minati, 2019a) of quasi-systems (Minati & Pessa, 2018), compatible with and condition for the occurring of processes of emergence, interesting in their equivalent multiplicity, that is, their coherence and uniqueness, implying irreducibility (e.g., non-complete analytical representations, non-zippability in formulas). The structural change particularly relates to cases when subsequent arrangements, organizations, and interactions differ so that they, as in complex collective and chaotic behaviors, are at least not linearly deducible from the previous one.

The emergence of multiple coherence levels in this case preserves the general identity of the complex collective and chaotic behaviors. It is similar to coherent sequences of unique phase transitions-like changes from uniqueness.

On the other side, structural dynamics may occur in less radical ways, allowing linear correspondences as considered in Sect. 1, such as for some social systems, e.g., corporate, clientele, and school behaviors.

In Sect. 4, we consider the case when multiple systems are not considered to be occurring sequentially but are both simultaneous or *pending*. Pending systems are intended as different versions of (almost) the same configurations of interacting composing elements waiting for external events to occur, prevail, come out, collapse, or be decided, i.e., authorized by the external events to take place. This is in conceptual correspondence with the quantum collapse of several superimposed eigenstates into a single one.

We conclude that, in order to be general, General Systems Theory should be reformulated in terms of multiple systems, avoiding simplified abstractions intended to represent general, approximated cases, such as statistical variables; gray systems-characterized by incompleteness in information of such uncertain systems (Javanmardi et al., 2020; Liu & Yang, 2012); fuzzy systems; and sloppy systems with models with many parameters of fitting (such models are intended as sloppy because poorly constrained and conditioned due to difficulties in using the experimental data) as in (Transtrum et al., 2015).

A more appropriate reformulation should introduce models theoretically considering multiplicities, incompleteness, and quasi-ness that are irreducible, non-simplifiable, and, in turn, become non-multiple, complete models. *A General Systems Theory not considering such aspects has implicit seeds of reductionism.*

2 Systems of Multiple Composing Elements, Multiple Interactions, and Emergent Multiple Systems

In psychology, for instance, the concept of multiple systems was introduced several years ago (Tulving, 1985) with the introduction of multiple-memory-systems models to account for findings of stochastic independence of direct priming effects and recognition memory effects.

In the introduction, we mentioned multiple systems (Minati & Pessa, 2006, pp. 110–116) as given by multiplicities of changes in composing elements and structures of interaction. This is a matter of structural dynamics of interest when it is suitable to maintain the emergence of systemic properties. Cases of structural dynamics occur, for instance, when there are processes of:

(a) replacing composing elements losing for any reason, e.g., wear and tear, properties such as the ability to (regularly) react.

Other cases relate to processes of interaction when separately or together in any combination there is the occurrence of:

(b) acquisition of a structure, i.e., change from a non-structured configuration to a structured one,

(c) change in structure, i.e., from one structure to another,
(d) loss of structure, i.e., change from a structured configuration to a non-structured one,
(e) combinations of structures.

An intervention on a multiple system causes corresponding variations in its other non-equivalent multiple versions; this enables manipulation in social systems, e.g., when acting on one allows influence on another. It relates to the multiple roles of their composing agents acting as simultaneous components of different systems such as classrooms, families, markets, the system of customers in a shop, temporary system communities such as passengers, members of vehicular traffic, and working places. *Languages, models, and stereotypes introduced in one version reappear in another*. However, it is a way to maintain general coherence.

Simple examples of simultaneous multiple systems include cases such as devices for which the cooperative and multiple roles of sensors constitute, at the same time, regulatory and safety systems; and values in electrical networks simultaneously establishing control and regulatory systems. It is a matter of multiple effects, meanings, and usages of the same information. Other cases occur when elements simultaneously belong to multiple networks, where the same nodes may belong to different networks simultaneously (Nicosia et al., 2013), as in the case of the brain and the Internet.

We notice how the simultaneous roles of constituent elements in multiple systems are of interest because they reciprocally reproduce and reuse the same changes that are identically transposed from the context of one system to the other. However, this identical reusing is temporary enough to allow the composing elements and interactions to adapt to the other system. This process should be carefully considered since it is composed of the dynamics of continuous re-usages having different conclusive endings, such as slowing the dynamics of re-usages until they reach convergent or divergent situations, allowing the multiple systems to eventually reach multiple equilibrium states, or diverging situations, allowing only temporary, local partial stabilities or full instabilities that induce randomness (Minati, 2021b, 2021c, 2022b).

At this point, we mention the multiplicity considered by the DYnamic uSAge of Models (DYSAM) when the same problem or phenomenon (not necessarily systems) can be represented in different ways and modeled using different approaches (Minati & Pessa, 2018, pp. 201–204; Minati & Pessa, 2006, pp. 64–85). The conceptual background of DYSAM includes ensemble learning, evolutionary game theory, and the Bayesian method, statistical approaches based on a "continuous exploration" of the events occurring within the phenomenon to be modeled. The multiplicity of DYSAM is methodological. It applies when problems have different, non-equivalent aspects that should be approached in different ways. For instance, a problem may currently be psychological, physiological, and social; a business problem may be financial, organizational, managerial, and related to marketing; a social problem may be political, economic, and sociological; a phenomenon can be thermodynamic and chemical; a mathematical problem can have algebraic and geometric solutions. It also applies when the multiplicity is related to structural dynamics rather than different aspects of the same system. Multidisciplinary approaches are insufficient and should be, rather, interdisciplinary when an aspect cannot be considered while ignoring the others or the multiple effects of interventions. This is the case when dealing with complex systems such as collective behaviors and simultaneously using models considering, for instance, density, energy, metrical aspects, topological aspects, and scale-free correlations.

2.1 Non-multiple System Modeling

The classic model of non-multiple systems is given by the following system of ordinary differential equations:

$$\begin{cases} dQ_1/dt = f_1(Q_1, Q_2, \ldots, Q_n) \\ dQ_2/dt = f_2(Q_1, Q_2, \ldots, Q_n) \\ \ldots \\ dQ_n/dt = f_n(Q_1, Q_2, \ldots, Q_n) \end{cases} \tag{1}$$

This system is intended given by n fixed elements p_i $(i = 1, 2, \ldots, n)$ for which there exist some measurements Q_i $(i = 1, 2, \ldots, n)$ and interacting through fixed rules of interaction f_n. Change of any measure Q_i, therefore is a function of all other Q's. The change of any Q, entails change of all other measures and of the system as a whole.

The instantaneous values of $Q_1, Q_2, \ldots,$ and Q_n are intended to specify the state of the system. The assumption is that model (1) generally applies to any generic system and to its changes over time.

However, Bertalanffy himself wrote "Such a definition of "system" is, of course, by no means general. It abstracts from spatial and temporal conditions, which would be expressed by partial differential equations." (Bertalanffy, 1968, p. 56).

2.2 Multiple System Modeling

The concept of multiple systems may be correspondingly analytically represented through structural variations, for instance when f_n changes in $f_{n,t}$ and Eq. 1 becomes time-dependent:

$$\begin{cases} dQ_1/dt = f_{1,t}(Q_1, Q_2, \ldots, Q_n) \\ dQ_2/dt = f_{2,t}(Q_1, Q_2, \ldots, Q_n) \\ \ldots \\ dQ_n/dt = f_{n,t}(Q_1, Q_2, \ldots, Q_n) \end{cases} \tag{2}$$

Similarly, state variables Q_n become time-dependent since the n elements p_i are variable in the sense that one element may quit for any reason, disappear, reappear, or be replaced, and new elements may be introduced. A schematic conceptual example is given in the following system of equations.

$$\begin{cases} \mathrm{d}Q_1/\mathrm{d}t = f_{1,t}(Q_1, Q_2, \ldots, Q_n) \\ \mathrm{d}Q_2/\mathrm{d}t = f_{2,t}(Q_1, Q_2, \ldots, Q_n) \\ \mathrm{d}Q_3/\mathrm{d}t = f_{3,t}(Q_1, 0, \ldots, Q_n) \\ \ldots \\ \mathrm{d}Q_n/\mathrm{d}t = f_{n,t}(Q_1, Q_2, \ldots, Q_{n+1}) \\ \mathrm{d}Q_{n+1}/\mathrm{d}t = f_{n+1,t}(Q_1, Q_2, \ldots, Q_{n+1}) \end{cases} \tag{3}$$

The case is further generalized by considering multiple systems characterized by almost partial simultaneities involving common composing elements and interactions (Minati & Pessa, 2006, pp. 123–128).

However, the dynamics of changing and replacing should be limited so that the system does not simply degenerate into configurations of independently, irregularly interacting elements losing coherence and then the system's nature, such as Brownian motion. Furthermore, the analytical representations used, while useful to represent the concept (Minati, 2022b, pp. 103–106), are absolutely useless and not applicable to represent real cases, intractable if not for those having forms of repetitiveness and very limited structural dynamics.

Effective mathematical models should be based on the usage, for instance, of combinations of analytical and non-analytic computational approaches such as networks, artificial neural networks (ANN) and networks changing levels and number of nodes, such as Recurrent Neural Networks (RNN), see, for instance (Bianchi et al., 2017; Newman et al., 2006), and nature-inspired computational approaches, see, for instance (Brabazon et al., 2015; Mac Lennan, 2004; Yang et al., 2013), and meta-structures, i.e., structures and interaction structures between clusters having variable components represented by mesoscopic variables (Minati & Licata, 2012, 2013, 2015; Minati, 2012a, 2012b).

The subsequent step is to consider the multiplicity, indefiniteness, and incompleteness of structural dynamics not as an analytical complication but as a necessary characteristic of emergent complex systems that are analytically intractable.

We discuss how considering the model of a system *in Eq.* 1 *as general is significant, to all effects, as a fact of reductionism that is effectively only applicable in cases of reduced or no structural dynamics, such as in non-complex systems, such as typical single-function mechanical and electronic artificial systems.*

3 The Unavoidable Multiplicity of the Emergence of Complex Systems

The concepts of "theoretical incompleteness" and "quasi-ness" have been introduced in conceptual contrast with the one of completeness. In particular:

(a) theoretical incompleteness (Longo, 2011, 2019; Minati, 2016; Minati et al., 2019) intended as incompleteness in principle, i.e., intrinsically non-complete-able, such as for uncertainty principles in physics; the complementarity principle, which states that some physical objects have complementary properties that cannot be observed or measured simultaneously (corpuscular and wave aspects of a particle); computational non-decidability, i.e., non-Turing computability (unavailability of a procedure));
(b) quasi-ness (Minati & Pessa, 2018) when there are non-equivalent representations of the same system, a system is not always a system, not always the same system, and not only a system; all of these necessary ingredients for the establishment of emergence processes are provided by multiple subsequent and predominantly coherent systemic variations. The limits of formal absolute validity of physical laws (Feynman, 1967, pp. 45 and 47) have been represented analytically in different ways, such as assuming statistical and probabilistic levels, ranges of validity, fuzziness as for fuzzy systems, and usage of imaginary numbers and complex variables such as in quantum physics.

The incompleteness of multiplicity, on the other hand, is identified as necessary for establishing the processes of emergence that would otherwise be blocked in the bud, reduced to predictable and computable, turning off multiplicities of equivalences and symmetry-breaking, real engines of emergence established by multiple variables, infinities.

In contrast *to "complete-able incompleteness,"* "theoretical *incompleteness"* is intended as *a* property *that can* leave a system in a state of permanent indefiniteness *caused* by multiplicities of equivalent and non-equivalent combined elements and interactions *and para-metrical* and timing values *that are primarily* always different.

The status of a system that is theoretically incomplete and theoretically logically open (Minati et al., 1996, 1998) *is conceivable as a network of equivalent possibilities from which configurations endowed with partial and dynamic coherence are selected, for instance, as a result of environmental perturbations and the occurrence of random predominance due to fluctuations*. This is the case for processes of emergence. Theoretical incompleteness may be intended as the place where phenomena are incomplete enough to allow the emergence of multiple dynamical coherence. Theoretical incompleteness is necessary, even if it is not sufficient for emergence processes to occur. The validity of appropriate constraints and forms of repetitiveness probably gives the condition of sufficiency.

Furthermore, theoretical incompleteness allows for the consideration of structurally variable quasi-systems (Minati, 2018, 2019a, 2021b), which should not be confused with fuzzy systems, where fuzziness deals with properties of belonging under the assumptions of structural representation stability and invariability. Quasi-systems are very realistic, highly compatible representations of real systems.

We consider, for instance, processes whose coherence effects are variable, *de-localized*, tentative, and even failing (i.e., not convergent or losing convergence, such as failed *healing* or only set diseases), subject to decay and to recovery; processes reaching predominance or not, occurring in variable amounts and percentages over time.

In models and simulations of the science of complexity, such theoretical incompleteness should be addressed, deemed insignificant, or approximated by well-defined and well-tractable statistical and non-linear analytical representations.

However, in the phenomenology of complexity (and not in its representation), such incompleteness is necessary and not (or almost not) completely zippable in formal representations, as considered in (Licata & Minati, 2016).

3.1 *Multiplicity of Emergent Multiple Systems*

In processes of emergence, for instance, multiplicity does not refer to well-defined systems, even if elements and interactions are partitioned as in Sect. 1. Rather, multiplicity relates to quasi- and incomplete systems that can maintain levels of coherence sufficient to make the *nature of system* manifest, i.e., the ability to acquire properties that composing elements do not have and even their linearly organized combinations (such an ability relates to the consistence and robustness of the emergent system's nature).

The dynamics of emergent collective systems refer to the complexity of continuously transforming quasi-systems while maintaining sufficient levels of coherence or quasi-coherence such as long-range correlations and variable, recoverable in losses, predominance of collective multiple partial remote synchronizations (Gambuzza et al., 2013; Minati et al., 2022a).

The emergence of (a) collective systems such as anthills, flocks, herds, schools, swarms, and vehicular traffic, and (b) ecosystems from continuous dynamics of roles, where the environment is also very difficult to identify, which in turn would be a continuously emerging complementary multiple system and multiple incomplete quasi-networks where nodes and links vary in intensity and occurrence (on–off) as in the brain, are typical examples of cases.

At this point, we must mention the theoretically central role of the so-called *weak forces* (Minati, 2021c). Without going into details, we can characterize forces as "weak" when they;

- are of "low" intensity, such as less than the minimum of a suitable percentage of all forces involved at the moment in the phenomenon under study;
- they have local ranges of influence involving very few adjacent composing elements.

Consequently, they have a range of influence and intensity insufficient to force changes in the prevalent *strong* forces active in the entire system unless all active forces are weak compared to the resulting total forces, such as flock speed.

The role and effectiveness of weak forces are in their ability to be decisive in breaking equivalences; equilibria; starting collapses; in deciding in situations of metastability and in catastrophe theory (Gilmore, 1993); and setting decisive initial conditions as in deterministic chaos. Furthermore, partial, simplified, and collective weak interactions may be assumed as tentative *initial conditions* for a self-establishing, quasi-convergent process of emergence that has the potential to dominate or fail. Examples include *spontaneous synchronizations*, unstable *remote synchronizations*, and tentative *long-range synchronizations* until a single or multiple specific synchronizations become predominant and eventually iterate (Minati, 2015).

4 Pending Multiplicity

Finally, it is possible to consider another point of view on the processes of emergence, i.e., how emergence emerges (Minati, 2019b).

The point under study relates to the *sudden* occurrence of systemic transformations instead of soft, temporarily significant changes in structural dynamics. We mention two cases whose equivalence or non-equivalence should be studied, especially with regard to application consequences. The two approaches can coexist, but one can prevail over the other for reasons of effectiveness and appropriateness.

(a) In cases, for instance, of phase transitions and metastability the temporal immediacy is related to the fact that the conditions necessary for the stable existence of a structure valid at the time t_1 cease to be valid at the time t_2, and a new structure is replaced.
(b) It is also possible to consider the hypothetical presence of potential "superimposed" multiple systems related to pending, inactivated, implicit interactions waiting for suitable environmental conditions, e.g., energetic, to "collapse" in the activation of one of the pending multiple systems. This is conceptually similar to the collapsing mechanisms that, in Quantum Mechanics (QM), cause the wave function to collapse from a superposition of several eigenstates to a single eigenstate as a result of interaction with the external world. The situation considered here may be intended as a case of classic collapsing in contrast with quantum collapsing. This correspondence seems to be a form of potential validation of how the formalism of QM could be used in describing a system using classical physics and embedding it into a stochastic, noisy background (Minati & Pessa, 2006, pp. 279–281).

In the second case (b), it is a matter of finding a balance between the unpredictability of emergence and the admissibility of new properties, which are unpredictable but collapsible in current situations. As a network of ongoing interactions is established, a set of potential networks is correspondingly constituted at least as nearly compatible, equivalent, or symmetrically opposite, i.e., incompatible and in-equivalent, variations.

For instance, in the coherent, smooth, regular behaviors of crowds, there are implicit, catastrophic, destructive potential behavioral variations that are decided, selected, and activated, collapsing in the face of the occurrence of particular situations such as the irregular, critical increase of the crowding, the general stampede and the trampling of people with no escape route, and terrifying events such as explosions. A similar situation can be considered for variations in vehicular traffic. Furthermore, conceptually analogous cases may be considered for economic systems with opposite properties, whose alternance is nowadays regulated through the use, for instance, of financial regulatory interventions. Similar situations can be identified in critical variations in ecosystems. It is a matter of the dynamics of the prevalence of one possibility over another, such as in the inhomogeneous climate system.

However, we stress how the dynamics mentioned above are not reduced to possibilities that are linearly reducible from one to the other, i.e., a world of equivalences. Instead, the situation is significant when it involves non-equivalent alternatives, however admissible, that are incompatible and partially coexist, as in first-order phase transitions.

We mention how, in philosophy, the situation discussed in point (b) above and later elaborated on relates to so-called dispositionalism, which deals with dispositional properties (see, e.g., Bird, 2005, 2016).

Dispositionalism holds that the essence of a property *P* is completely constituted by the causal roles that *P* plays, with each causal role given in terms of its potential causes and effects. A common example is the fragility of things; for example, when a glass is sufficiently struck, it breaks. Brittleness is considered a dispositional property of glass that explains its breakage. Other examples of dispositional properties include solubility and flammability.

5 Conclusions

We considered different multiplicities of multiple systems, such as implying multiple, eventually variable, roles of elements and interactions; incompleteness and quasiness in processes of emergence; and potential "superimposed" multiple systems related to pending, inactivated, implicit interactions waiting for the occurring of suitable environmental conditions to "collapse." The latter case is considered related to the philosophical concept of dispositionalism.

The relationship between chaos, coherence of long-range interactions, and correlations should be better investigated.

Examples of open issues related to the possible or variable coherence or even incompatibility, local or global, of multiple systems and their possible adjustability and manageability. These abilities have important social manipulative effects, and conversely, their knowledge has important anti-manipulative defensive effects.

The General Systems Theory is general in the sense that the system model in Eq. 1 is supposed to apply to represent possible dynamics of any system, hypothesized having stable same elements p_i, measurements Q_j, and interaction mechanism given by f_n.

When considering theoretical quasi-multiplicity, a generalization of such general systems theory is intended as represented by the model of systems in Eq. 3.

The traditional General Systems Theory may be viewed as a subset of a more generalized version, Second Generation General System Theory (Minati et al., 2016, 2017). The reduction of the Second Generation General System Theory to the General Systems Theory is intended to be a case of reductionism with theoretical and practical consequences.

Only the study of the properties of models of systems in Eq. 3 should represent general system invariants rather than models of systems in Eq. 1.

Suitable approaches to modeling the multiplicity, theoretical incompleteness, and quasi-ness of the *Second Generation General System Theory* should be introduced, such as based on multiple approaches combining, for instance, fuzzy clustering (which determines the number of clusters) and fuzzy clustering (where each item can belong to more than one cluster) as in Miyamoto et al. (2008) and Tibshirani et al. (2001); multiple networking (Nicosia et al., 2013); remote synchronization (Minati et al., 2022a, 2022b); and meta-structures (Minati, 2009a, 2009b; Minati & Licata, 2015: Minati et al., 2013; Minati & Pessa, 2018, pp. 102–129) that have yet to be developed due to a lack of research funds.

References

Bertalanffy, L. (1968). *General system theory. Development, applications.* George Braziller.

Bianchi, F. M., Maiorino, E., Kampffmeyer, M. C., Rizzi, A., & Jenssen, R. (2017). *Recurrent neural networks for short-term load forecasting: An overview and comparative analysis.* Springer.

Bird, A. (2005). The dispositionalist conception of laws. *Foundations of Science, 10*, 353–370.

Bird, A. (2016). Overpowering: How the powers ontology has overreached itself. *Mind, 125*, 341–383.

Brabazon, A., O'Neill, M., & McGarraghy, S. (2015). *Natural computing algorithms.* Springer.

Feynman, R. (1967). *The character of physical law.* The MIT Press.

Gambuzza, L. V., Cardillo, A., Fiasconaro, A., Fortuna, L., Gómez-Gardenes, J., & Frasca, M. (2013). Analysis of remote synchronization in complex networks. *Chaos, 23*, 1–8.

Gilmore, R. (1993). *Catastrophe theory for scientists and engineers.* Dover.

Javanmardi, E., Liu, S., & Xie, N. (2020). Exploring the philosophical paradigm of grey systems theory as a postmodern theory. *Foundations of Science, 25*, 905–925. https://doi.org/10.1007/s10699-019-09640-5

Licata, I., & Minati, G. (2016). Emergence, computation and the freedom degree loss information principle in complex systems. *Foundations of Science, 21*(3), 1–19.

Liu, S., & Yang, Y. (2012). A brief introduction to grey systems theory. *Grey Systems: Theory and Application, 2*(2), 89–104. https://doi.org/10.1108/20439371211260081

Longo, G. (2011). Reflections on concrete incompleteness. *Philosophia Mathematica, 19*, 255–280.

Longo, G. (2019). Interfaces of incompleteness. In G. Minati, M. Abram, & E. Pessa (Eds.), *Systemics of incompleteness and quasi-systems* (pp. 3–55). Springer.

Mac Lennan, B. J. (2004). Natural computation and non-Turing models of computation. *Theoretical Computer Science, 317*(1–3), 115–145.

Minati, G. (2012a). Introduction to the meta-structures project: Prospective applications. *World Futures, 68*(8), 558–574.

Minati, L. (2015). Remote synchronization of amplitudes across an experimental ring of non-linear oscillators. *Chaos, 25*, 123107–123112.

Minati, G. (2016). Knowledge to manage the knowledge society: The concept of theoretical incompleteness. *Systems, 4*(3), 26. https://doi.org/10.3390/systems4030026

Minati, G. (2018). The non-systemic usages of systems as reductionism. Quasi-systems and quasi-systemics. *Systems, 6*(3), 28. https://doi.org/10.3390/systems6030028

Minati, G. (2019a). Non-classical systemics of quasi-coherence: From formal properties to representations of generative mechanisms. A conceptual introduction to a paradigm-shift. *Systems, 7*(4), 51. https://doi.org/10.3390/systems7040051

Minati, G. (Ed.). (2021a). *Multiplicity and interdisciplinarity—Essays in honor of Eliano Pessa*. Springer.

Minati, G. (2021c). What would I propose to Eliano to continue our collaboration: Validity regimes, edges, interfaces, and waves of complexity. In G. Minati (Ed.), *Multiplicity and interdisciplinarity—Essays in honor of Eliano Pessa* (pp. 93–119). Springer.

Minati, G., & Licata, I. (2012). Meta-structural properties in *collective behaviours. The International Journal of General Systems, 41*(3), 289–311.

Minati, G., & Pessa, E. (2006). *Collective beings*. Springer.

Minati, G., & Pessa, E. (2018). *From collective beings to quasi-systems*. Springer.

Minati, G., Penna, M. P., & Pessa, E. (1998). Thermodynamic and logical openness in general systems. *Systems Research and Behavioral Science, 15*(3), 131–145.

Minati, G., Abram, M., & Pessa, G. (Eds.). (2016). *Towards a post-bertalanffy systemics. Systemics of incompleteness and quasi-systems*. Springer.

Minati, G., Abram, M., & Pessa, G. (Eds.). (2019). *Systemics of incompleteness and quasi-systems*. Springer.

Minati, L., Bartels, J., Li, C., Frasca, M., & Ito, H. (2022a). Synchronization phenomena in dual-transistor spiking oscillators realized experimentally towards physical reservoirs. *Chaos Solitons and Fractals, 162*, 112415. https://doi.org/10.1016/j.chaos.2022.112415

Minati, G., & Licata, I. (2013). Emergence as mesoscopic coherence. *Systems, 1*(4), 50–65. http://www.mdpi.com/2079-8954/1/4/50

Minati, G., & Licata, I. (2015). Meta-structures as multidynamics systems approach. Some introductory outlines. *Journal on Systemics, Cybernetics and Informatics* (JSCI), *13*(4), 35–38. http://www.iiisci.org/journal/sci/issue.asp?is=ISS1504

Minati, G., Penna, M. P., & Pessa, E. (1996). Towards a general theory of logically open systems. In E. Pessa, M. P. Penna, & A. Montesanto (Eds.), *Proceedings of the 3rd Systems Science European Congress* (pp. 957–960). Kappa.

Minati, G., Licata, I., & Pessa, E. (2013). Meta-structures: The search of coherence in collective behaviours (without physics). In A. Graudenzi, G. Caravagna, G. Mauri, & M. Antoniotti (Eds.), *Proceedings Wivace 2013—Italian Workshop on Artificial Life and Evolutionary Computation (Wivace 2013), Electronic Proceedings in Theoretical Computer Science* (Vol. 130, pp. 35–42).

Minati, G., Pessa, E., & Licata, I. (Eds.) (2017). *Second generation general system theory: Perspectives in philosophy and approaches in complex systems*. MDPI. http://www.mdpi.com/books/pdfview/book/325

Minati, L., Bartels, J., Li, C., Frasca, M., & Ito, H. (2022b). *The whole in a part: Synchronizing chaos through a narrow slice of spectrum*. Tokyo Tech News. https://www.titech.ac.jp/english/news/2022/065324

Minati, G. (2009a). Steps of the meta-structures project to model general processes of emergence—Theoretical frameworks, the mesoscopic general vector, future lines of research and possible applications. Preprint at http://arxiv.org/ftp/arxiv/papers/0910/0910.1753.pdf

Minati, G. (2009b). The meta-structures project. Preprint at http://arxiv.org/ftp/arxiv/papers/0903/0903.0592.pdf

Minati, G. (2012b). Approaches and principles of the Meta-Structures Project: The mesoscopic dynamics. Notes for software and models designers. Preprint at https://arxiv.org/abs/1201.1475v2

Minati, G. (2019b). Phenomenological structural dynamics of emergence: An overview of how emergence emerges. In L. Ulivi (Ed.), *The systemic turn in human and natural sciences. A rock in the pond* (pp. 1–39). Springer.

Minati, G. (2021b). On modelling the structural quasiness of complex systems. *WSEAS Transactions on Systems and Control, 16*, 715–734. https://doi.org/10.37394/23203.2021.16.65

Minati, G. (2022a). Interactions and interaction mechanisms as generators of complex systems. *WSEAS Transactions on Proof, 2*, 96–107. https://doi.org/10.37394/232020.2022.2.12

Minati, G. (2022b). Multiplicity, logical openness, incompleteness, and quasi-ness as peculiar non-reductionist properties of complexity. In S. Wuppuluri, & I. Stewart (Eds.), *From electrons to elephants and elections: Saga of content and context* (pp. 151–173). Springer.

Miyamoto, S., Ichihashi, H., & Honda, K. (2008). *Algorithms for fuzzy clustering: Methods in C-means clustering with applications*. Springer.

Newman, M., Barabasi, A. -L., & Watts, D. J. (2006). *The structure and dynamics of networks*. Princeton University Press.

Nicosia, V., Bianconi, G., Latora, V., & Barthelemy, M. (2013). Growing multiplex networks. *Physical Review Letters, 111*(058701), 1–5.

Tibshirani, R., Walther, G., & Hastie, T. (2001). Estimating the number of data clusters via the gap statistic. *Journal of the Royal Statistical Society Series B: Statistical Methodology, 63*, 411–423.

Transtrum, M. K., Machta, B. B., Brown, K. S., Daniels, B. C., Myers, C. R., & Sethna, J. P. (2015). Perspective: Sloppiness and emergent theories in physics, biology, and beyond. *The Journal of Chemical Physics, 143*, 010901.

Tulving, E. (1985). How many memory systems are there? *American Psychologist, 40*, 385–398.

Yang, X.-S., Cui, Z., Xiao, R., Gandomi, A. H., & Karamanoglu, M. (Eds.). (2013). *Swarm intelligence and bio-inspired computation*. Elsevier.

Multiple Systems in Probability Theory Applications

Marco Giunti, Simone Pinna, and Fabrizia Giulia Garavaglia

Abstract In applying probability theory to specific problems, it is not uncommon that many random experiments are to be considered simultaneously, each of which has a particular probability space associated with it, and that the problem under consideration requires to find probabilities defined in a particular space from known probabilities defined in other spaces. In such cases, before probability theory can be applied, it is necessary to transfer the known probabilities from the spaces in which they are defined to the space in which the required probability is defined. This transfer can be justified only by using special principles independent of the axioms of probability theory, because the latter apply only to events defined in a single probability space. In this paper, by analyzing in depth a notable example of this kind of problem, we formulate and justify one such principle, the Principle of Equivalence.

1 Introduction

When applying probability theory to solve a specific problem, we first of all need to specify the formal systems we use to represent the random experiments and the associated events involved. Each of these formal systems is a probability space that consists of three elements: a sample space Ω, a sigma-algebra $\sigma(\Omega)$ built over Ω, and a probability function **p** defined on $\sigma(\Omega)$. Unless the problem at stake is extremely simple, there is more than one probability space involved and, to solve the problem, we need to determine the probabilities of the events in one space by means of the probabilities of the events in other spaces. This determination of probabilities through

M. Giunti (✉) · S. Pinna · F. G. Garavaglia
Dip. di Pedagogia, Psicologia, Filosofia, Università di Cagliari, Cagliari, Italy
e-mail: giunti@unica.it

S. Pinna
e-mail: simone.pinna@unica.it

F. G. Garavaglia
e-mail: fabriziag.garavaglia@unica.it

G. Minati and M. Pietronilla Penna (eds.), *Multiple Systems*, Contributions to Management Science, https://doi.org/10.1007/978-3-031-44685-6_2

multiple spaces can be made only by means of principles that are independent from the axioms of probability theory, for such axioms exclusively apply to events that belong to the same probability space. Thus, in applying probability theory to this type of problem, we need further principles for correlating multiple probability spaces.

To illustrate this situation, we consider a notable application of probability theory: Find the probability $\mathbf{P}(E_k^n)$ of getting k successes in a sequence of n Bernoulli trials. As it is well known, the solution to this problem is given by the binomial distribution formula

$$\mathbf{P}(E_k^n) = \binom{n}{k} p^k (1-p)^{n-k}$$

where p is the probability of success on an arbitrary trial. Nevertheless, we show that the usual procedure to reach this solution is flawed, for it does not take into account that more than one probability space is simultaneously involved in this problem. By considering this fact, we give a correct procedure and we make explicit, and justify, the Principle of Equivalence on which it is based. This principle is independent from the axioms of probability theory and, when used together with these axioms, allows us to apply probability theory to cases, such as Bernoulli trials, in which many probability spaces are simultaneously involved.

2 Bernoulli Trials

Consider a *Bernoulli experiment*, i.e., a random experiment t with only two possible results, 0 (failure) and 1 (success), which then give us the sample space of t:

$$\Omega := \{0, 1\}$$

we denote by ω a generic element of Ω, so $\omega = 0$ or $\omega = 1$.

Let us now consider a new random experiment

$$t^n := t_1, t_2, \ldots, t_n$$

t^n is called a *sequence of n Bernoulli experiments* just in case, for any i, t_i is a Bernoulli experiment.

Let $t^n = t_1, t_2, \ldots, t_n$ be a sequence of n Bernoulli experiments, and Ω_i be the sample space of t_i. Since t_i is a Bernoulli experiment, its possible results are 0 and 1, so that the sample space of each t_i is:

$$\Omega_i = \Omega$$

we denote by ω_i a generic element of Ω_i, so $\omega_i = 0$ or $\omega_i = 1$.

When $n > 1$, the possible results of t^n are all the n-tuples built out of $\Omega = \{0, 1\}$. Thus, the sample space of t^n is the n-th Cartesian product of Ω:

$$\Omega^n = \Omega_1 \times \cdots \times \Omega_n = \Omega \times \cdots \times \Omega$$

we denote by ω^n a generic element of Ω^n. When $n = 1, t^1 = t_1$, therefore the sample space of t^1 is $\Omega^1 := \Omega$, and we denote by ω^1 a generic element of Ω^1.

For each $\omega^n \in \Omega^n$ and every i such that $1 \leq i \leq n$, we denote by ω_i^n the i-th component of ω^n; thus, $\omega_i^n = 0$ or $\omega_i^n = 1$. Also keep in mind that, by definition, ω_i^n is a possible result of t_i, so that $\omega_i^n = \omega_i$.

Assuming that, for each Bernoulli experiment t_i, the probability of a success occurring is the same and that the result of t_i does not depend on the result of any set of other experiments, we want to determine the probability that the result of t^n has exactly k successes.

To state this problem more precisely, let S_i, for each i and n $(1 \leq i \leq n)$, be the event «the result of t_i is 1», that is to say:

$$S_i := \{\omega_i \in \Omega_i : \omega_i = 1\} = \{1\}$$

Also, for each $k (0 \leq k \leq n)$, let E_k^n be the event «the result of t^n has exactly k successes», that is to say:

$$E_k^n := \left\{\omega^n \in \Omega^n : \omega^n \text{ has exactly } k \text{ components equal to } 1\right\}$$

As the sample spaces Ω_i and Ω^n are all discrete (in fact, they are finite), the corresponding sigma-algebras are taken to be identical to their power sets:

$$\text{for all } i, \sigma(\Omega_i) := \mathcal{P}(\Omega_i) \text{ and } \sigma\left(\Omega^n\right) := \mathcal{P}\left(\Omega^n\right)$$

For each i, let $\mathbf{p}_i$ the probability function defined on the sigma-algebra $\sigma(\Omega_i)$, and $\mathbf{P}$ the probability function defined on the sigma-algebra $\sigma(\Omega^n)$. We can now state the problem as follows:

Problem (What is the probability of getting k successes in a sequence of n Bernoulli trials?)

Assume that:

(i) for all i, $\mathbf{p}_i(S_i) = p$, where p is a fixed real number such that $0 \leq p \leq 1$;
(ii) for all i, the result of t_i does not depend on the result of any set of other experiments.

Given these two hypotheses, determine $\mathbf{P}\left(E_k^n\right)$.

If a sequence of n Bernoulli experiments $t^n = t_1, t_2, \ldots, t_n$ satisfies both conditions (i) and (ii), it is called a *sequence of n Bernoulli trials*, and each t_i is called a

Bernoulli trial. With this definition, our problem can be formulated more synthetically as: What is the probability of getting k successes in a sequence of n Bernoulli trials?

3 The Usual Solution of the Problem

Before presenting the usual way in which this problem is solved, we note that it has two hypotheses. The first one states that the probability of getting a success is the same for each experiment t_i. Formally, this means that the n probability spaces for the n experiments are identical. It is important to emphasize that this hypothesis must be stated explicitly, because it is not implied by the fact that all the n experiments $t_1, t_2, \ldots, t_n$ have the same sample space $\Omega_i = \Omega$ and thus, being Ω discrete, also the same sigma-algebra $\sigma(\Omega_i) = \mathcal{P}(\Omega)$, namely the power set of Ω. In fact, the probability function $\mathbf{p}_i$ defined on $\sigma(\Omega_i)$ might be different for different experiments. The *hypothesis of equiprobability* of success through different experiments eliminates this possibility.

The second hypothesis states that the result of each experiment t_i does not depend on the result of any set of other experiments. The intuitive meaning of this *hypothesis of independence* is clear but, from the point of view of probability theory, it must be expressed more precisely in terms of the independence of appropriate families of *events*. We will see below how this can be done.

Finally, we note that the problem requires to find $\mathbf{P}\big(E_k^n\big)$, that is, the probability of an event E_k^n that belongs to the sigma-algebra $\sigma(\Omega^n) = \mathcal{P}(\Omega^n) \neq \mathcal{P}(\Omega) = \sigma(\Omega_i)$. It is thus requested to find a probability in the probability space $(\Omega^n, \mathcal{P}(\Omega^n), \mathbf{P})$, by only knowing, for each i, the probability $\mathbf{p}_i(S_i)$ in the space $(\Omega_i, \mathcal{P}(\Omega_i), \mathbf{p}_i) = (\Omega, \mathcal{P}(\Omega), \mathbf{p})$. This means that, despite its apparent simplicity, the problem posed is not at all trivial, because it requires us, so to speak, to push the limits of applicability of probability theory which, as such, only allows us to calculate probabilities within a single space.

Let us now present the usual solution of the problem, as can be found in excellent university-level textbooks (see for example: Feller, 1968, pp. 146–148; Hogg et al., 2020, pp. 169–170).

1. ***The cardinality of*** $\boldsymbol{E_k^n}$

We note first that the n trials of the random experiment t^n constitute a set of n distinct elements $T = \{t_1, t_2, \ldots, t_n\}$ and that there is a one-to-one correspondence f between E_k^n and the set of all subsets of k elements of T, where $f(\omega^n) := \big\{t_i \in T : \omega_i^n = 1\big\}$, for all $\omega^n \in E_k^n$. The number of possible results in E_k^n is therefore:

$$\big|E_k^n\big| = \binom{n}{k} = \frac{n!}{k!(n-k)!}$$

For each $\omega^n \in \Omega^n$, consider the corresponding singleton (or elementary) event $\{\omega^n\}$. Obviously, all singleton events are incompatible with each other and furthermore each event E_k^n is the union of all singleton events $\{\omega^n\}$ such that $\omega^n \in E_k^n$. Thus, $\mathbf{P}(E_k^n)$ can be obtained as the sum of the probabilities of such singleton events:

$$\mathbf{P}(E_k^n) = \sum_{\omega^n \in E_k^n} \mathbf{P}(\{\omega^n\})$$

Therefore, it is now a matter of finding out how we can determine the probability of any singleton event.

2. ***The probability of any singleton event*** $\{\omega^n\}$ ***such that*** $\omega^n \in E_k^n$

For any k, we consider an arbitrary possible result $\omega^n \in E_k^n$. Being $\omega_i^n = \omega_i$ for every $i (1 \le i \le n)$, the singleton event $\{\omega^n\}$ corresponds to the occurrence of all events

$$\{\omega_1\}, \{\omega_2\}, \ldots, \{\omega_n\}$$

and it can thus be thought as their conjunction. For the hypothesis of equiprobability, each of these events has probability:

$$\mathbf{p}_i(\{\omega_i\}) = \begin{cases} p & \text{if } \omega_i = 1 \\ 1 - p & \text{if } \omega_i = 0 \end{cases}$$

For the hypothesis of independence, the result of each experiment t_i is independent of the result of any set of other experiments, therefore these probabilities multiply and thus:

$$(*) \quad \mathbf{P}(\{\omega^n\}) = \prod_{i=1}^{n} \mathbf{p}_i(\{\omega_i\})$$

Moreover, being $\omega^n \in E_k^n$, there are exactly k events $\{\omega_i\}$ such that $\omega_i = 1$. So, by adopting the convention $0^0 := 1$, we get:

$$\mathbf{P}(\{\omega^n\}) = p^k (1 - p)^{n-k}$$

N.B.

The convention $0^0 := 1$ allows the inclusion of the two cases $p = 0$ and $p = 1$, which otherwise should be excluded. In fact, in the first case, if $k = 0$ the above formula becomes:

$$\mathbf{P}(\{\omega^n\}) = 0^0 1^n$$

in the second case, if $k = n$,

$$\mathbf{P}(\{\omega^n\}) = 1^n 0^0$$

Without that convention, 0^0 would be an indeterminate form and therefore, in both cases, $\mathbf{P}(\{\omega^n\})$ would be indeterminate.

3. ***The binomial distribution***

Finally, recall (see above) that $\binom{n}{k}$ is the number of possible results $\omega^n \in E_k^n$, and that $\mathbf{P}(E_k^n)$ is equal to the sum of the probabilities of all singleton events $\{\omega^n\}$ such that $\omega^n \in E_k^n$. From this and the equation just found, we thus obtain the well-known formula of the binomial distribution:

$$\mathbf{P}(E_k^n) = \binom{n}{k} p^k (1-p)^{n-k}$$

4 Criticism of the Solution Procedure

Although the solution found for our problem is undoubtedly correct, the procedure by which we reached it has a dubious step that is worth highlighting and discussing in more detail.

This step, which we have marked with an asterisk $(*)$, occurs in the argument that allows us to determine, for an arbitrary $\omega^n \in E_k^n$, the probability $\mathbf{P}(\{\omega^n\})$ of the corresponding singleton event $\{\omega^n\}$.

According to this step, the probability $\mathbf{P}(\{\omega^n\})$ of the event $\{\omega^n\}$ would be identical to the product of the probabilities $\mathbf{p}_i(\{\omega_i\})$ of all the events $\{\omega_i\}$ and this identity would be justified by the fact that, by hypothesis, the result of each experiment t_i is independent of the result of each set of other experiments. But this in fact means that in order to get $(*)$,

(i) we interpret the independence hypothesis as the independence of the n events $\{\omega_1\}, \{\omega_2\}, \ldots, \{\omega_n\}$, and
(ii) we implicitly identify the event $\{\omega^n\}$ with their intersection.

In fact, only with these assumptions can we get $(*)$ from the independence hypothesis and the definition of independence. However, once assumptions (i) and (ii) are made explicit, we quickly understand that they are not tenable.

First, the n events $\{\omega_1\}, \{\omega_2\}, \ldots, \{\omega_n\}$ cannot be independent. Since $\omega^n \in E_k^n$ is arbitrary, let us suppose $\{\omega_1\} = \{1\}$, $\{\omega_2\} = \{1\}$ and $0 < p < 1$; then, $\{\omega_1\} \cap \{\omega_2\} = \{1\}$ and so $\mathbf{p}(\{\omega_1\} \cap \{\omega_2\}) = p$. Furthermore, since $\{\omega_1\} = \{1\} = \{\omega_2\}$, $\mathbf{p}(\{\omega_1\}) = \mathbf{p}(\{\omega_2\}) = p$. Hence, $\mathbf{p}(\{\omega_1\})\mathbf{p}(\{\omega_2\}) < p$, so that $\mathbf{p}(\{\omega_1\} \cap \{\omega_2\}) \neq \mathbf{p}(\{\omega_1\})\mathbf{p}(\{\omega_2\})$; therefore, $\{\omega_1\}, \{\omega_2\}, \ldots, \{\omega_n\}$ are not independent.

Second, the event $\{\omega^n\}$ cannot be the intersection of $\{\omega_1\}, \{\omega_2\}, \ldots, \{\omega_n\}$. In fact, such an intersection is empty if at least two $\{\omega_i\}$ are different; otherwise, depending on whether all the $\{\omega_i\}$ are equal to $\{0\}$ or $\{1\}$, the intersection is equal to $\{0\}$ or $\{1\}$, respectively. On the contrary, $\{\omega^n\}$ is not empty and is not identical to either $\{0\}$ or $\{1\}$, because ω^n is a n-tuple of zeros and ones.

Looking more closely at assumptions (i) and (ii), we realize that their untenability depends on the fact that the event $\{\omega^n\}$ on the one hand, and the corresponding events $\{\omega_1\}, \{\omega_2\}, \ldots, \{\omega_n\}$ on the other, belong to different sigma-algebras associated with different random experiments. In fact, $\{\omega^n\}$ belongs to the sigma-algebra $\sigma(\Omega^n) = \mathcal{P}(\Omega^n)$ consisting of all subsets of the sample space Ω^n, which is related to the random experiment t^n; on the other hand, each $\{\omega_i\}$ belongs to the sigma-algebra $\sigma(\Omega_i) = \mathcal{P}(\Omega)$ consisting of all subsets of the sample space $\Omega_i = \Omega$, which is related to the random experiment t_i.

It is therefore not at all surprising that $\{\omega^n\}$ cannot be formally expressed as the intersection of the $\{\omega_i\}$s, even though, from an intuitive point of view, $\{\omega^n\}$ can be thought of as their conjunction. Similarly, neither can the independence hypothesis be formally expressed as the independence of the $\{\omega_i\}$s.

5 How to Solve the Difficulty

To overcome this impasse, we must therefore find appropriate replacements of the events $\{\omega_1\}, \{\omega_2\}, \ldots, \{\omega_n\}$, which we denote by $R_1^n(\omega^n), R_2^n(\omega^n), \ldots, R_n^n(\omega^n)$, for which the following conditions hold:

(a) each $R_i^n(\omega^n)$ belongs to the same sigma-algebra as $\{\omega^n\}$;
(b) $\{\omega^n\}$ is identical to their intersection;
(c) the independence hypothesis is expressible as their independence;
(d) each $R_i^n(\omega^n)$ has the same probability as $\{\omega_i\}$, that is:

$$\mathbf{P}\big(R_i^n(\omega^n)\big) = \mathbf{p}_i(\{\omega_i\}) = \begin{cases} p & \text{if } \omega_i = 1 \\ 1-p & \text{if } \omega_i = 0 \end{cases}$$

To this end, for each $n \geq 1$, for each i such that $1 \leq i \leq n$, let S_i^n be the event «the result of t^n has a 1 (success) in position i» and F_i^n be the event «the result of t^n has a 0 (failure) in position i», that is to say:

$$S_i^n := \left\{\omega^n \in \Omega^n : \omega_i^n = 1\right\}$$

$$F_i^n := \left\{\omega^n \in \Omega^n : \omega_i^n = 0\right\}$$

We note that F_i^n is nothing more than the complement, or the negation, of S_i^n, that is:

$$F_i^n = \overline{S_i^n}$$

Now, for an arbitrary possible result $\omega^n \in \Omega^n$, we can define:

$$R_i^n(\omega^n) := \begin{cases} S_i^n & \text{if } \omega_i^n = 1, \\ \overline{S_i^n} & \text{if } \omega_i^n = 0. \end{cases}$$

We verify below that $R_1^n(\omega^n)$, $R_2^n(\omega^n), \ldots, R_n^n(\omega^n)$ satisfy the two conditions (a) and (b). We then show how also conditions (c) and (d) can be met.

Verification of condition (a)

By definition, each $R_i^n(\omega^n)$ is a subset of Ω^n and therefore belongs to the same sigma-algebra as ω^n.

Verification of condition (b)

By definition, for each i, $\omega^n \in R_i^n(\omega^n)$ and, in addition, no other possible result belongs to each $R_i^n(\omega^n)$. Therefore, $\{\omega^n\} = R_1^n(\omega^n) \cap R_2^n(\omega^n) \cap \cdots \cap R_n^n(\omega^n)$.

How to meet condition (c)

According to the statement of our problem (see Sect. 2), the independence hypothesis states that, for all i, the result of t_i does not depend on the result of any set of other experiments. We noted earlier (see Sect. 3, par. 2) that the intuitive meaning of this hypothesis is clear but, from the point of view of probability theory, it should be expressed more precisely in terms of the independence of appropriate families of *events*. That is, it should be rephrased as follows: for any family $R_1^n, \ldots, R_n^n$ of n events such that each R_i^n expresses a possible result of experiment t_i, such events are independent. By definition, for all ω^n, each event $R_i^n(\omega^n)$ of the family $R_1^n(\omega^n), \ldots, R_n^n(\omega^n)$ expresses exactly one possible result of t_i, i.e., 1 if $R_i^n(\omega^n) = S_i^n$, 0 if $R_i^n(\omega^n) = F_i^n$. We can then formulate the independence hypothesis in the following way:

For all ω^n, $R_1^n(\omega^n), \ldots, R_n^n(\omega^n)$is a family of independent events

In fact, we can simplify this formulation by recalling that, given any family of independent events, if we replace some events with their complements, we always get another family of independent events (Berger et al., 2021, Proposizione 1.70). We note that, once an arbitrary ω_*^n is fixed, the family $R_1^n(\omega^n), \ldots, R_n^n(\omega^n)$ corresponding to any other ω^n is obtained from the family $R_1^n(\omega_*^n), \ldots, R_n^n(\omega_*^n)$ by replacing some $R_i^n(\omega_*^n)$ with their complements. Therefore, it is sufficient to assume only the independence of the family $R_1^n(\omega_*^n), \ldots, R_n^n(\omega_*^n)$, because that of all others follows from it. For convenience, the family $R_1^n(\omega_*^n), \ldots, R_n^n(\omega_*^n)$ is usually chosen such that, for all i, $R_i^n(\omega_*^n) = S_i^n$. We will conform to this convention, thus reformulating the independence hypothesis as follows:

S_1^n, ..., S_n^n is a family of independent events

How to meet condition (d)

We now want to show that, for each ω^n and every i,

$$(**)\quad \mathbf{P}\big(R_i^n(\omega^n)\big) = \mathbf{p}_i(\{\omega_i\}) = \begin{cases} p & \text{if } \omega_i = 1 \\ 1-p & \text{if } \omega_i = 0 \end{cases}$$

We first note that equation $(**)$ identifies the probabilities of two events that belong to different sigma-algebras. In fact, $R_i^n(\omega^n)$ is an event belonging to the sigma-algebra $\sigma(\Omega^n) = \mathcal{P}(\Omega^n)$, while $\{\omega_i\}$ belongs to the sigma-algebra $\sigma(\Omega_i) = \mathcal{P}(\Omega)$. Consequently, this identity cannot be a consequence of the axioms of probability theory alone, because these axioms hold only for events belonging to a single sigma-algebra.

However, although the two events $R_i^n(\omega^n)$ and $\{\omega_i\}$ belong to different sigma-algebras, there is a clear sense in which they can be said to be *equivalent*. For, according to the definitions of the two events, $R_i^n(\omega^n)$ occurs if and only if $\{\omega_i\}$ occurs. To make this point clearer, let us first define precisely the notion of occurrence of an event.

Let t be a generic random experiment to which a sample space Ω with sigma-algebra $\sigma(\Omega)$ is associated, and let $r(t)$ be one and only one of the elements of Ω, called *the result of the random experiment* t. Let E be any event belonging to $\sigma(\Omega)$.

Definition (Occurrence of an event)

We say that the event E *occurs*: iff $r(t) \in E$.

Using the previous definition, we can now define the relation of *equivalence between two arbitrary events*, which not necessarily belong to the same sigma-algebra.

Let $\sigma(\Omega_1)$, $\sigma(\Omega_2)$ be two arbitrary sigma-algebras, and E_1, E_2 be any two events belonging to $\sigma(\Omega_1)$, $\sigma(\Omega_2)$, respectively.

Definition (Equivalence between two events)

E_1 is *equivalent to* E_2: iff E_1 occurs if and only if E_2 occurs or, in other words, if the occurrence of one implies the occurrence of the other and vice versa.

It is now easy to show that $R_i^n(\omega^n)$ and $\{\omega_i\}$ are equivalent. If $R_i^n(\omega^n)$ occurs, $r(t^n) \in R_i^n(\omega^n)$ and then, by the definition of $R_i^n(\omega^n)$, if $\omega_i^n = 1$, $r(t_i) = 1$; if $\omega_i^n = 0$, $r(t_i) = 0$. But $\omega_i^n = \omega_i$, so, in both cases, $r(t_i) \in \{\omega_i\}$ and therefore $\{\omega_i\}$ occurs. Conversely, if $\{\omega_i\}$ occurs, $r(t_i) \in \{\omega_i\}$ and so if $\omega_i = 1$, $r(t_i) = 1$; if $\omega_i = 0$, $r(t_i) = 0$. Since $\omega_i^n = \omega_i$, in the first case, $R_i^n(\omega^n) = S_i^n$ and so $r(t^n) \in R_i^n(\omega^n)$; in the second case, $R_i^n(\omega^n) = F_i^n$ and thus $r(t^n) \in R_i^n(\omega^n)$. Therefore, in both cases, $R_i^n(\omega^n)$ occurs.

Note that, in the special case of two events belonging to the same sigma-algebra, equivalence between the two events reduces to their identity and thus equivalence is not a particularly significant relation. The importance of this relation appears only in the case of two equivalent events belonging to different sigma-algebras, because *such events must always have the same probability*. In fact, if this were not the case, we would be assigning different probabilities to two events that always occur (or do not occur) jointly, which is absurd. In other words, we are thereby asserting the following Principle of Equivalence:

Principle of Equivalence

If two events are equivalent, they have the same probability.

In the case of two events belonging to the same sigma-algebra, the Principle of Equivalence is a trivial consequence of the fact that two equivalent events are identical. But, when the two events belong to different sigma-algebras, the principle gains its full relevance, because it allows us to transfer the same probability from one sigma-algebra to another. Note also that, in the case of events belonging to different sigma-algebras, the Principle of Equivalence cannot be a consequence of the axioms of probability theory, because these axioms apply only to events belonging to a single sigma-algebra. The Principle of Equivalence is thus an independent principle that, used in conjunction with the usual axioms, allows us to apply probability theory to problems in which it is necessary to consider more than one sigma-algebra simultaneously. The determination of the probability of getting k successes in a sequence of n Bernoulli trials is one such problem.

In fact, it is precisely this principle that allows us to derive condition (d). Since $R_i^n(\omega^n)$ and $\{\omega_i\}$ are equivalent, ($**$) follows from the Principle of Equivalence; therefore, condition (d) holds.

Having seen how all four conditions (a), (b), (c), (d) are satisfied, we can now reformulate the second part of the solution to our problem correctly.

2'. The probability of any singleton event $\{\omega^n\}$ such that $\omega^n \in E_k^n$—Correct procedure

For any k, we consider an arbitrary possible result $\omega^n \in E_k^n$. For condition (b):

$$\{\omega^n\} = R_1^n(\omega^n) \cap R_2^n(\omega^n) \cap \cdots \cap R_n^n(\omega^n)$$

For condition (d):

$$\mathbf{P}(R_i^n(\omega^n)) = \mathbf{p}_i(\{\omega_i\}) = \begin{cases} p & \text{if } \omega_i = 1 \\ 1-p & \text{if } \omega_i = 0 \end{cases}$$

For condition (c), S_1^n, ..., S_n^n is a family of independent events and therefore so is $R_1^n(\omega^n), \ldots, R_n^n(\omega^n)$. Hence, from the previous two equations, it follows:

$$\mathbf{P}(\{\omega^n\}) = \prod_{i=1}^{n} \mathbf{p}_i(\{\omega_i\})$$

In addition, since $\omega^n \in E_k^n$, there are exactly k events $\{\omega_i\}$ such that $\omega_i = 1$. Therefore, adopting the convention $0^0 := 1$, we obtain:

$$\mathbf{P}(\{\omega^n\}) = p^k(1-p)^{n-k}$$

6 Concluding Remarks

When it comes to applying probability theory to specific problems that are even moderately complex, it is not uncommon that many random experiments are to be considered simultaneously, each of which has a particular probability space associated with it, and that the problem under consideration requires to find probabilities defined in a particular space from known probabilities defined in other spaces. In such cases, before probability theory can be applied, it is necessary to *transfer* the known probabilities from the spaces in which they are defined to the space in which the required probability is defined. This transfer can be justified only by using special principles independent of the axioms of probability theory, because the latter apply only to events defined in a single probability space.

In this paper, by analyzing in depth a notable example of such a problem, the determination of the probability of getting k successes in a sequence of n Bernoulli trials, we were able to formulate and justify one of these principles, the Principle of Equivalence, and showed how it allows us to arrive at the solution of the problem correctly. We maintain that the method used to solve this particular problem can be a useful model for dealing with new problems of this kind. But we also believe that it has its own intrinsic philosophical value, because it shows that even the understanding and justification of a well-known result, if properly problematized, can be deepened and improved.

Appendix

For the reader's convenience, we recall here some important definitions used in this paper.

Power set

Let X an arbitrary set. *The power set of* X is the set of all subsets of X, and is denoted by $\mathcal{P}(X)$.

Factorial

For each $n \in \mathbb{Z}^{\geq 0}$, *the factorial of* n, denoted by $n!$, is defined by recursion as follows:

$$0! = 1$$
$$(n+1)! = (n+1)n!$$

Binomial coefficient

For each $n, k \in \mathbb{Z}^{\geq 0}$, *the binomial coefficient* $\binom{n}{k}$ gives us the number of all subsets of k elements of a set of n elements. Equivalently, we can say that $\binom{n}{k}$ is the number of ways in which k distinct elements can be chosen among n distinct elements, regardless of the order.

The binomial coefficient is defined as follows:

$$\binom{n}{k} := \begin{cases} \frac{n!}{k!(n-k)!} & \text{if } 0 \leq k \leq n \\ 0 & \text{if } 0 \leq n < k \end{cases}$$

Family of independent events

Let $(E_i)_{i \in I}$ an arbitrary family of events in a probability space $(\Omega, \sigma(\Omega), \mathbf{p})$.

$(E_i)_{i \in I}$ is a family of *independent* events: iff for every finite subset $J \subseteq I$, if J has at least 2 elements, it holds:

$$\mathbf{p}\left(\bigcap_{j \in J} E_j\right) = \prod_{j \in J} \mathbf{p}(E_j)$$

References

Berger, Q., Caravenna, F., & Dai Pra, P. (2021). *Probabilità*. II edizione. Springer Italia.

Feller, W. (1968). *An introduction to probability theory and its applications* (Vol. 1, 3rd ed.). Wiley.

Hogg, R. W., McKean, J. W., & Craig, A. T. (2020). *Introduction to mathematical statistics* (8th ed.). Pearson Education Limited.

Logical Open Systems as Oracles

Ignazio Licata

Abstract In this work, we propose a line of research on logical open systems interpreted as Turing oracles coupled to other systems, suggesting that the resulting metasystem produces new results and computational emergences.

1 Introduction

Much of the progress in information technology is aimed at expanding the resources of computing systems in terms of processing time/speed and memory capacity. In particular, in order to bypass the Von Neumann's "bottleneck"—computation step by step—there has been designed massively parallel architectures that distribute processing among multiple nodes of interconnected computation/memory. In some cases, as an alternative to the current silicon chips, different materials have been used to make the logic gates; it is the case of bio-chips or the optical computers. In none of these cases, however, the fundamental theory of computation that derives from the formidable work of Alan Turing on computable numbers is radically questioned (Turing, 1937). Here, we want to propose some highly speculative considerations on the possibility of defining new models of computation, equivalent or even more powerful than the one offered by Turing-computability. This is possible by considering the metasystem that is obtained by coupling a suitable logical open system (LOS) with a system within which the undecidable problem is defined. This is equivalent to establishing constraints that modify the "semantics" of the system (Licata, 2008; Licata & Minati, 2017; Minati & Licata, 2012; Minati et al., 1998).

I. Licata (✉)
Institute for Scientific Methodology (ISEM), Via Ugo la Malfa 153, I-90146 Palermo, Italy
e-mail: ignaziolicata3@gmail.com

School of Advanced International Studies on Theoretical and Nonlinear Methodologies of Physics, I-70124 Bari, Italy

International Institute for Applicable Mathematics and Information Sciences (IIAMIS), B.M. Birla Science Centre, Adarsh Nagar, Hyderabad 500463, India

G. Minati and M. Pietronilla Penna (eds.), *Multiple Systems*, Contributions to Management Science, https://doi.org/10.1007/978-3-031-44685-6_3

2 The Turing Limit

Alan Turing developed his famous Turing Machine (TM) considering what a human being actually does when he/she performs a calculation. Otherwise from what one would be led to think today, the problem did not have a purely applicative vocation, but was rather linked to axiomatization program by D. Hilbert and the limits highlighted in 1931 by K. Gödel with his famous theorems of undecidability and incompleteness: there is no deductive inferential system capable to answer as regards the truth or falsehood of every proposition constructable through enumerable recursive functions. In practice, a TM provides the correct and universal mathematical context for the general definition of algorithm, in accordance with the Church-Turing thesis: TMs are a class of computationally universal automata (Hodel, 2013).

Beyond the various formalisms (register machines, Post production systems, Lambda-Calculus Church, Chomsky's Grammars, etc.), TM remains the brightest and clearest conceptual model of computation, and can be described as an infinite strip on which a finite state automaton can read, write and manipulate a finite number of symbols operating according to a set of rules applied in discrete time, step by step. These few observations suggest that the notion of Turing-computability (T-comp) originates in a highly formalized environment and is targeted to the syntactical analysis of information (Hodel, 2013). Indeed, the concept of information has much broader aim, which we can understand by referring to physical systems. In nature, there is no information that is not associated with some form of discrete and/or continuous matter-energy; any physical or biological system can be considered a processor of information, in which the value of some input quantity comes transformed into output by the internal dynamics of the system. In this sense, we can say that every physical system "computes", and other aspects of information must be taken into consideration: dynamics (the space–time distribution of information) and semantics (the value that information has for the system, a particularly important issue in the study of biological systems). Therefore, we can ask if new scenarios related to the study of chaotic, quantum and biological systems suggest new models of computation. TMs are characterized by some fundamental limitations closely related to the K. Gödel's theorems, and globally indicated as Turing's limit. It is possible to prove that TMs are a countably infinite set (i.e., they can be counted), while the set of all functions that can be constructed has the power of the continuum, i.e., they are an uncountable infinite set. It is quite intuitive to understand that this result is connected to the discreet aspect of the T-comp, and to the way of enumerating TM. In other words, the degree of "infinity" of TMs is lower than that of the set of constructible functions, and therefore, there exist non-Turing computable functions (non-recursive functions). A classic example is given by the halting problem: given an algorithm and a string, it is possible to prove that there is no TM able to decide whether the computation process will end or not, i.e., whether or not the algorithm will produce that string. Computers never know when it is time to stop! It can be useful to see the result in terms of formal languages: with the tools of the language, it is impossible to establish whether a certain expression belongs to language or not, then:

there are recursively enumerable, but not recursive, languages (theorem on recursive languages).

Intuitively, it is possible to connect this limitation to a "rigidity" of TMs, of languages and formal systems. Indeed, it is possible to show than to overcome a single undecidability problem—but there will always be infinitely, according to the universality of Gödel-Turing results!—new axioms must be added to the system, i.e., the structure of the system must be modified. We are now so accustomed to living with the obviousness of this result that we fail to appreciate the game of implications to which the violation of this limitation would lead.

3 Oracles

We consider a hybrid structure of the type (TM, O), where the symbol O indicates an ideal device called oracle. In its broadest sense, an oracle can be considered as a set of decision procedures capable of constraining the activity of the TM. It could be a table of values, or more general a "black-box". What matters here is to realize the kind of limitations of a TM. By definition, an oracle is able to overcome the limits of a TM, e.g., by providing the algorithm of a non-recursive set with respect to TM, and thus solving the halting problem. In practice, access to the oracle would make a hybrid machine capable of coping with classes of problems that no algorithmic system can solve. Another field where oracles have shown a theoretical function is in the classification of "tractable" or "intractable" problems, that is, solvable in polynomial or exponential time in relation to the n-dimension of the problem. The study of classes P and NP is one of the great open problems of mathematics, but some convincing arguments have shown that the problems of the "NP-complete" type (such as the problems of optimization, from the "traveling salesman" to the folding protein), can be reduced to "tractable" problems with polynomial time algorithms with the help of oracles. However, the same line of reasoning has shown that the type of solution strictly depends on the features of the adopted oracle, and therefore does not have a universal but rather a problem-based character.

It is possible to shed new light on the features of the TMs and the hybrid machines using the theory of logical open systems (LOS). As known, all the physical systems can be classified using three main categories: (1) conservative systems of information, which conserve energy; (2) information destroyer systems, subject to dissipation; (3) information amplifier systems, where emergence and self-organization phenomena occur in polynomial or exponential time. This third class includes logical open systems, described through a formal logical open model, such as a non-Turing-Computable system, for which it is impossible to establish an actual procedure for realizing a priori what information is relevant to the description of the system. This is because the information is provided to the system in a non-algorithmic (random) way and—above all—because the game of interrelationships between system and environment is particularly complex and leads to continuous structural reorganization of the system that cannot be made explicit through a simple recursive procedure. This class includes nonlinear and context-dependent systems, such as swarms or collective beings, and neural networks (Minati & Pessa, 2006). These systems show a type of emergence called intrinsic or observational, not attributable to a computational production. You can classify the LOS in a hierarchy of orders of complexity, and prove that it is impossible to describe a logical open system by means of a single formal model, i.e., a TM, and that a high degree system of logical openness can only be approximated by another with an openness of degree n, with n interpretable as the number of constraints which formalize the relations between system and environment. On the other part, a TM can be identified with a typical logically closed system (LCS), whose global architecture does not change, or in any case it can be always derived from an effective procedure, which is an essential reason for historical chessmates of the "strong" AI. So, we come to the essential point: *a TM is a logical closed system, an oracle is a logical open system* (with a very high degree of logical openness).

Even if an ideal oracle does not exist and is the result of the speculative genius of A. Turing, it is possible to have devices with "oracle" capabilities that correspond to varying degrees of logical openness, such as neural networks or autonomous agents of distributed AI. The study of these systems suggests a new analogical super-Turing computational paradigm (Syropoulos, 2007; Ord & Kieu, 2005; Kieu, 2003, 2004).

4 Turing Super and Subsystems

The study of logical open systems with "oracular" capabilities finds its inspiration in a large number of physical systems. This corresponds to the obvious consideration that the problems of "incompleteness" and "undecidability" that afflict our formal systems do not exist in nature. The T-comp is only one of the possible ways we can adopt to describe the flow of information in natural and artificial processes. A feature that should be emphasized, in addition to the nonlinearity, sensitivity to the context, evolutionary and self-organizational multi-level capacities is the continuity which sets the "degree of infinity" of oracle systems to the power of the continuum. For example, the configurations of a neural network can be parameterized through a real parameter connected to the weights of the net. In general, the study of continuous formal systems, still in the germinal state, allowed us to identify the significant features of natural computation with respect to the TMs (see Table 1).

But in practice, what physical systems can actually be used in order to make super-Turing capable devices? In this sense, there are some interesting, all quite recent, "indications" which we can only mention here. There are classes of dynamical systems that can solve particular formulations of the halting problem. Recurrent Neural Networks (RNR) to synchronous modification and with real weights can obtain very good performance on NP-complete problems, such as asynchronous DAI systems, and also some quantum algorithms show that they can solve problems of exponential difficulty for classical systems of computation, in polynomial

Table 1 Significant features of natural computation with respect to Turing machines

Turing computability	Natural computation-formal continuous systems
Programmability; the process terminates after a finite number of steps; undecidability problems, halting problem	Self-organizing evolutionary skills; the process does not end, it modifies its aims in relation to the context
Discrete information; binary logic; syntactic distribution of information	Continuous information; multiple-value logic; continuous patterns in space–time
Deterministic process; computational emergence	Random elements; intrinsic emergence
Accurate data; defined output	"Fuzzy" data; adaptive optimization strategies
Serial/linear	Parallel/nonlinear

times. The same indications come from DNA-computing, where the information suitably "engraved" on the DNA is processed by enzymatic chains. Finally, even chaos computing, heir of the control techniques of chaotic systems, is entering a mature phase, full of great possibilities on optimization problems (Aoun, 2016; Berthiaume & Brassard, 1994; Calude & Pavlov, 2002; Currin et al., 2017; Deutsch & Josza, 1992; Kieu, 2002, 2004; Licata, 2007, 2012; Ord, 2002; Ord & Kieu, 2005; Powell, 2022; Siegelmann, 2012).

It is therefore possible to imagine a technology of neuro-fuzzy, chaotic and quantum oracles. The price to pay is universality. In all the mentioned cases, in fact, it is possible to provide counter-examples in which the performance is compromised by small variations of the involved parameters, in correspondence with the emerging critical thresholds of computing capacity. Beyond the particular cases, an intuitive consideration exists that can help us understand this coexistence of super and sub-Turing capabilities. Logical open systems present phenomena of intrinsic emergency and therefore are subject to a high degree of logical unpredictability. The "oracular" computation therefore appears as a form of "dedicated" computation, built on the particular problem and the characteristics of the system. The thing should not surprise us, because we know well a device of this type, developed by the evolution in millennia of "trials and errors": the human mind.

5 The Riemann Oracle

Generally, the notion of logical open system is used in reference to physical and biological systems, but there is no problem in extending it to formal structures. Let us consider a mathematical structure A made up of a set of elements on which relationships are defined. We can extend or constrain A with another structure A′ of elements and relations without changing the initial structure. We will say that A′ has a greater logical openness than A because it is characterized by a greater number of constraints. Let us now use these concepts in the specific case of a classic number theory problem (Hardy & Wright, 2008).

It is known that the research on Hilbert's tenth problem led to the establishment of a Diophantine equation capable of generating prime numbers in a non-exclusive way (Matiyasevich, 1993; Jones et al., 1976). Since then, several polynomials of this type have been produced, improved in the number of variables or in the degree, but it has never been possible to obtain a Diophantine equation capable of skimming only the primes. The interest in this approach has therefore progressively decreased in favor of the Riemann hypothesis, which affirms the belonging of the primes to a critical strip (Mazur & Stein, 2016). We note that these are in fact two different approaches: distribution formulas generate primes, the Riemann Hypothesis is a statement about the order of primes. For aesthetic reasons, mathematicians work as if the Riemann Hypothesis had already been proved, and in fact significant even if not definitive clues accumulate year by year (e.g., Cassettari et al., 2022; Tamburini & Licata, 2021), while the laws of distribution are affected by decidability problems (Minati, 2019). In this situation, it may be interesting to consider the two approaches as equals, as different aspects of the mystery of the primes, without any preconceived idea of deducibility of one from the other.

It is known that an equation of the Matyasevich type is equivalent to a Turing machine (MT) and therefore to a cellular automaton of class IV (Langton, 1990; Wolfram, 2002). As it is well known, it is a typical stochasticity that we could define as "weak", characterized, roughly speaking, by increasingly rarefied "bursts" of primes (see, e.g., Lemke & Soundararajan, 2016). Equations of the Matyasevich type produce not only primes, but also negative numbers which correspond to the trivial zeros of the Z function. This would appear to be a strong limitation, but it reveals actually an interesting clue. Indeed, we know that the Riemann hypothesis is equivalent to stating that the absolute value of the difference between the enumerative function of the primes and the integral logarithm $l\ \pi(x) - \mathrm{li}(n)\ l$ constrains the reciprocal oscillation of the two functions on an order O of magnitude going as $O\sqrt{n}\ \ln(n)$ (Derbyshire, 2003; du Sautoy, 2003).

If we also consider trivial zeros, using the same line of reasoning, we can use the Möbius function $\mu(n)$ to show that the difference between the number of k from 1 to n for which $\mu(k) = -1$, and the number of k from 1 to n for which $\mu(k) = +1$ is $O\sqrt{n}$, of the order of square root of n, i.e., it is bounded by a constant on the order of the square root of n (Chaitin, 2004). In other words, once again the Riemann hypothesis turns out to be an oracle on the distribution of primes which reveals a particular randomness we called "bursts" of primes, ultimately depending on the number of prime factors of n. The apparent amount of "garbage" produced by Matyasevich-like Diophantine equations is actually the indication of a very refined sieve that selects coin tosses and whose trend is given by $O\sqrt{n}$.

These ideas, in + seminal form and strongly conjectural, on the role of oracle of Riemann's function Z do not exclude its full demonstrability in the future, but suggest a perspective on the effective relationship between the Riemann hypothesis and the "natural" organization of the primes. From this point of view, a convergence between the Riemann hypothesis and the Diophantine machines is possible (Fodden, 2011), and perhaps delving deeper into the roots of undecidability (Chaitin, 1990).

References

Aoun, M. A. (2016). Advances in three hypercomputation models. *EJTP, 13*(36), 169–182.

Berthiaume, A., & Brassard, G. (1994). Oracle quantum computing. *Journal of Modern Optics, XLI.*

Calude, C. S., & Pavlov, B. (2002). Coins, quantum measurements, and Turing's barrier. *Quantum Information Processing, 1*, 107–127.

Cassettari, D., Mussardo, G., & Trombettoni, A. (2022). Holographic realization of the prime number quantum potential. *PNAS Nexus, 2*(1), 279.

Chaitin, G. J. (1990). *Information, randomness and incompleteness*. World Scientific.

Chaitin, G. J. (2004). Thoughts on the Riemann hypothesis. *Mathematical Intelligencer, 26*, 4–7.

Currin, A., Korovin, K., Ababi, M., Roper, K., Kell, D. B., Day, P. J., & King, R. D. (2017). Computing exponentially faster: Implementing a non-deterministic universal Turing machine using DNA. *Journal of the Royal Society Interface, 14*, 128.

Derbyshire, J. (2003). *Prime obsession: Berhhard Riemann and the greatest unsolved problem in mathematics*. Joseph Henry Press.

Deutsch, D., & Jozsa, R. (1992). Rapid solution of problems by quantum computation. *Proceedings of Royal Society*, A439.

Du Sautoy, M. (2003). *The music of the primes*. Harper.

Fodden, B. (2011). Diophantine equations and the generalized Riemann hypothesis. *Journal of Number Theory, 131*, 1672–1690.

Hardy, G. H., & Wright, E. M. (2008). *An introduction to the theory of numbers*.

Hodel, R. (2013). *An introduction to mathematical logic*. Dover.

Jones, J. P., Sato D., Wada, H., & Wiens, D. (1976). Diophantine representation of the set of prime numbers. *American Mathematical Monthly, 83*, 449–464.

Kieu, T. D. (2002). Quantum hypercomputation. *Minds and Machines, 12*, 541–561.

Kieu, R. D. (2003). Computing the non-computable. *Contemporary Physics, 44*, 1.

Kieu, T. D. (2004). Hypercomputation with quantum adiabatic processes. *Theoretical Computer Science, 317*(1–3), 93–104.

Langton, C. (1990). Computation at the edge of chaos: Phase transitions and emergent computation. *Physica D: Nonlinear Phenomena, 42*(1–3), 12–37.

Lemke Olivier, R. J., & Soundararajan, K. (2016). Unexpected biases in the distribution of consecutive primes. *PNAS, 113*(31).

Licata, I. (2007). Effective physical processes and active information in quantum computing. *Quantum Biosystems, 1*(1), 51–65.

Licata, I. (2008). Logical openness in cognitive models. *Epistemologia, XXXI*, 177–192.

Licata, I. (2012). Beyond Turing: Hypercomputation and quantum morphogenesis. *Asia Pacific Mathematics Newsletter, 2*(3), 20–24.

Licata, I., & Minati, G. (2017). Emergence, computation and the freedom degree loss information principle in complex systems. *Foundations of Science, 22*, 863–881.

Matyasevich, Y. (1993). *Hilbert's 10th problem*. MIT Press.

Mazur, B., & Stein, W. (2016). *Prime numbers and the Riemann hypothesis*. Cambridge Univ. Press.

Minati, G. (2019). A conceptual proposal on the undecidability of the distribution law of prime numbers and theoretical consequences. *Ratio Mathematica, 37.*

Minati, G., & Licata, I. (2012). Meta-structural properties in collective behaviors. *International Journal of General Systems, 41*, 289–311.

Minati, G., & Pessa, E. (2006). *Collective beings*. Springer.

Minati, G., Penna, M. P., & Pessa, E. (1998). Thermodynamic and logical openness in general systems. *Systems Research and Behavioral Science, 15*(3), 131–145.
Ord, T. (2002). Hypercomputation: Computing more than the Turing machine. arXiv:math/0209332v1
Ord, T., & Kieu, T. D. (2005). The diagonal method and hypercomputation. *The British Journal for the Philosophy of Science, 56*(1).
Powell, A. (2022). Computation and hypercomputation. *Mathematics, 10*(6), 997.
Siegelmann, H. T. (2012). *Neural networks and analog computation: Beyond the Turing limit*. Birkhauser.
Syropoulois, A. (2007). *Hypercomputation: Computing beyond the Church-Turing barrier*. Springer.
Tamburini, F., & Licata, I. (2021). Majorana quanta, string scattering, curved spacetimes and the Riemann hypothesis. *Physica Scripta, 96*, 125276.
Turing, A. M. (1937). On computable numbers, with an application to the entscheidungsproblem. *Proceedings of the London Mathematical Society, 42*(1), 230–65.
Wolfram, S. (2002). *A new kind of science*. Wolfram Media.

A Theoretical Model for EEG Interpretation

Pier Luigi Marconi, Rosamaria Scognamiglio, Maria Lidia Mascia, Rachele Conti, Caterina Marconi, and Maria Petronilla Penna

Abstract How systems can be functionally integrated is of particular interest in neurosciences, while explaining the underlying bioelectrical phenomena of consciousness. Recent reviews have proposed the cross frequency phase-amplitude coupling (CfM) as a way by which the local brain activities (at high frequency) are integrated in large functional brain networks (by low frequency). Large brain networks are needed for consciousness and for strong cognitive tasks as well as they may be disrupted in pathological conditions as schizophrenia or emotional dysregulation. We have made a math model of high-frequency wave activities (by 9 dipoles) integrated by low-frequency modulators. In this model, "integration" is intended as *phase coherence* and *activity inter correlation* between dipoles. The highest integration was found to be linked to specific combination of modulation waves and dipole frequencies and phase differences. These results lead to the hypothesis that integrating the brain networks is not only necessary a low-frequency modulation in the Delta/Theta band, but also a predisposition of dipoles getting an harmonic resonance with that waves; this interpretation highlights the possibility that brain integration needs the past presence of harmonic resonance between dipoles and modulators, in relation with the personal history of parental functioning and neural development.

Keywords Complex systems integration · Emergent properties · Consciousness · Cross frequency modulation · Inter dipole correlation · Brain networks · Functional integration

---

P. L. Marconi (✉) · R. Scognamiglio · C. Marconi
Artemis Neurosciences, Rome, Italy
e-mail: pierluigi.marconi@artemisonline.org

M. L. Mascia · R. Conti · M. P. Penna
Department of Pedagogy, Psychology, Philosophy, University of Cagliari, Cagliari, Italy

C. Marconi
Sapienza University of Rome, Rome, Italy

G. Minati and M. Pietronilla Penna (eds.), *Multiple Systems*, Contributions to Management Science, https://doi.org/10.1007/978-3-031-44685-6_4

1 Background

A topic of interest for systemics is to study how the components of a system can be integrated each other. On functioning of highly integrated complex systems, the emergence of new properties (emergent properties) arises, such as, e.g., consciousness in neural systems. Emergent properties are not only new ways of perceiving the system by the outer world and new ways of describing its behavior as a whole, but they represent also phenomena with the ability to top-down influence the underlying system. This action causes a modification of the a priori probabilities of the interactions between nodes according to new matrices of probabilities that, in turn, reinforce the persistence of the emergent properties. In the neurosciences, we have a very particular emergent property that has been studied by several scholars and researchers for hundred years: consciousness and awareness. Till today, the underlying mechanisms are still largely unclear. While at the origins, the study was fundamentally clinical (as Jean-Martin Charcot and Sigmund Freud did), today the study of consciousness is performed mainly by mean of observation of Electroencephalographic (EEG) signals (Modolo et al., 2020). The electromagnetic signals produced by the cortex sum each other, generating very complex and non-homogeneous final fields. The EEG signal is a composite signal, constituted by several dipoles from different spatial sources added together at each detection point. In addition to spatial differences, the EEG signal also has different frequency components. We can distinguish different classes of these frequency bands. The first two bands concern very low frequencies ranging from 2 to 8 Hz (Delta band and Theta band), followed by a second class which includes the band from 8 to 13 Hz (Alpha band). From 14 to 32 Hz, we have a fourth band (Beta) and then we have the fastest electrical activities (from 32 to 140 Hz and beyond). A different meaning has been attributed to each of these bands. They can contribute to the detected EEG signal in different percentage; the different aspects of the EEG, due to different frequency classes' contributions, and their distribution over the scalp are associated to different clinical observation. In general, the signals at higher frequencies tend to maintain a more local distribution, while signals at lower frequencies tend to propagate more widely, also in relation to the fact that the brain is a dispersive system and the speed of propagation varies in relation to the frequency (Modolo et al., 2020). The lowest frequencies appear to be the fastest and are best propagated by the myelinated pathways of the white matter, representing a rapid way of interconnection between different brain areas. If the EEG has an intrinsic complexity, also the underlying systemic structure (the brain) has a fairly complex anatomy, with different brain areas involved in different functions. However, this compartmentalization, on an anatomical and functional basis, does not represent the real functional organization of the brain. In operative conditions, various different components are activated in different functional patterns, according to the macro-task on duty. It has recently been observed that brain activity tends, according to the task, to organize itself in functional structures defined as Brain Networks (BNets) (Bressler & Menon, 2010; Raichle, 2011). Currently, the main BNets identified are three:

1. the Default Mode Network activated above all in rest conditions and when the person carries out an internal thought activity, having access to his own memory content;
2. the Salience Network activated in the presence of a new stimulus, through which an initial value is attributed to the external stimulus (neutral, positive or negative) and therefore the appropriate cortical areas are activated for its more in-depth evaluation;
3. the Central Executive Network activated when an operational task must be carried out, structuring a problem, defining a consequent an executive plan and therefore monitoring the implementation process and the real effects.

It has been observed that in the presence of switching from one BNet to another one, the basic electrical activity is related to low-frequency waves which can be interpreted as functional BNet reorganizers. An example of this reorganization activity is the electrical reaction to external stimuli. These event related "waves" have historically been classified according to the direction of the observed variation and the time required to reach the maximum peak (positive or negative); some authors, however, have recently hypothesized that rather than primary waves they can be interpreted as the effect of modulation waves of the Delta and Theta band increasing and synchronizing the electrical activity in certain areas of the cortex, related to the characteristics of the stimuli (Klimesch et al., 2004). In a study published in 2007, by L. Melloni and coll., found that the activity of consciousness correlates with an increase in coherent activity in the high Gamma band (60–80 Hz), with the first observable elements around 200 ms, and a highly evident increase of phase synchrony at 800 ms. What are the electrophysiological elements at the basis of this phase coherence has interested the researchers in the last 20 years. The reorganization of the brain in coherent electrical patterns occurs, according to some authors, by modulating the amplitude of the higher frequency activity by the phase of the electrical activities of lower frequency band (Cross Frequency Modulation—CfM) (Lisman & Jensen, 2013; van Wijk et al., 2015; Marconi et al., 2016). This CfM has been observed in many studies, both in physiological and pathological conditions, representing a possible way by which the same neural system is functionally reorganized according to the task to be performed (Moran & Hong, 2011; Axmacher et al., 2010; Allen et al., 2011). The results published by Melloni in 2007 can be interpreted as the effect of cross frequency modulation of a large BNets, which leads to that complex systemic integration needed for rising consciousness. This process of systemic integration, while appearing increasingly important and increasingly involved in brain functional dynamics, at the same time still does not appear sufficiently clarified in all its preconditions and functional characteristics. The low-frequency modulating waves seem to originate from deep cerebral structures, as the basal nuclei or the thalamus, but what anatomical and functional implications derive from it is still not clear.

2 Objective of the Study

The aim of this work is to develop a mathematical model for simulating cerebral electromagnetic activity, in which local high-frequency activities are modulated by the phase of lower frequency activities.

Through this simulation model, it is expected to observe the implications and the interactive elements linked to the effects of the phase-amplitude CfM processes. In fact, it is possible that some fundamental variables at the basis of the process have not been already identified due to the fact that, in natural physiological conditions, these variables are already fixed by the process of psycho-neurodevelopment.

3 Methods

3.1 The Math Modeling

A very simple model has been built based on the interaction of 9 dipoles, whose activity is assumed to be recorded at a certain distance from a single probe.

Each dipole represents a coherent source of EEG signal, due to the mass effect of many nearby cortical pyramidal neurons involved in very similar tasks.

The model presented and tested here includes different math formulas linked together, representing the activation and deactivation of dipole (affecting the amplitude A, of the signal), and its time course, following a wave pattern with a characteristic oscillation ω, phase φ, time t at a point (x, y) in the space. The phase of each dipole is modulated by the phase of a family of low-frequency waves (from 2 to 32 Hz). The phase of these modulating waves is affected by the delay due to the propagation in a dispersive medium. The effect of dispersive medium is represented by the *wave number* $k = (2\pi f)/v$, with f = *frequency* of the modulation wave and v = *propagation speed* computed by a logarithmic function related the oscillation frequency.

The 9 dipoles are arranged on the perimeter of a circular space and described by specific values of the two spatial axes x, y.

The low-frequency modulation activities are generated by structures inside this space and they are also spatially described by specific values of x and y. Low-frequency activity generators are divided into 3 classes. A first class (oscillators at 2, 4, 8 Hz) is represented by three generators of global organization activity and are connected, modulating, both to the 9 dipoles and to the generators of the second class. The second class (oscillators at 16, 20, 24, 28, 32 Hz) is represented by 5 activity generators of local organization, which are connected to different subgroups of dipoles whose activity they modulate. The third class (12 Hz) is represented by a single generator, set centrally in the system, connected to all the 9 dipoles.

The final EEG signal is computed as detected by a probe E, located at point $p(x, y)$.

The model allows you to easily reconfigure the connections of the generators between them and with the different dipoles. The simulation can be activated with different patterns of modulating generators, as can be hypothesized to occur in vivo as a function of the mental task (BNets).

3.2 *The Simulation Trials*

Ten trials were performed for each of eight different activation patterns of modulators (all, none, 2–8 Hz, 12 Hz, 16–32 Hz and 3 further mixed patterns).

The oscillation of each dipole was randomly set on each trial between 80 and 110 Hz with an initial phase shift randomly placed at $\varphi = 1 \pm 0.1\ \omega t$.

The performance of the system was evaluated as

(a) effect of the modulation on the correlation between amplitudes of dipole pairs (*inter dipole correlation-IDC*)
(b) reduction of the phase difference between them with respect to that expected as a function of the simple original oscillation of the dipole (*ME*).

A cluster analysis was performed to detect the effective configuration of modulators in relation with IDC and ME.

The correlation between clusters or the eight original patterns and the IDC or ME was performed to detect the most effective.

Three different factorial analyses were performed on three groups of parameters

1. the activation of the nine modulators all together plus the frequency of the dipoles torques;
2. all the parameters assessing the phase and the frequency of the two dipole with and without modulation;
3. all the factors extracted by the previous analysis.

The criteria of *eigen value* > 1 1 was adopted to identify the significant factors, and an *oblimin rotation* was done to identify the best load of each parameter on each factor.

An analysis of variance was performed on each extracted factor in relation to the computed clusters.

On the basis of the results of previous analysis, five single shot simulations were done, computing on each one a spectral analysis of the composite signal at the probe by Fourier Fast Transform.

4 Results

In total, 800 simulations were performed on 36 different combinations of dipoles, for a total of 2880 observations.

The actually average oscillation frequency of the two dipoles of each pair was 94.7 ± 8.7 Hz for the first dipole and 94.8 ± 8.6 Hz for the second dipole (Table 1). On average, the difference in the oscillation frequency of the two dipoles of the pair was 9.7 ± 7.1 Hz.

The composite signal at the probe is within the High Gamma band, as the signal frequency is about 95 Hz on average.

After modulation, the phase difference between the two dipoles decreased from 0.020 ± 0.457 ωt to −0.008 ± 0.992 ωt, with a reduction in the difference (ME) of −0.593 ± 0.723 ωt.

The absolute correlation value (aIDC) of the activity between the two dipoles was $R = 0.0156$.

Table 1 Descriptive statistics

		Min	Max	Mean	Standard deviation
Frequency	Dipole 1 frequency	80	110	94.71	8.697
	Dipole 2 frequency	80	110	94.84	8.634
	Frequency difference between dipoles	−29.33	28.50	0.13	12.06
	Absolute frequency difference between dipoles	0.00	29.33	9.73	7.12
Phase	Phase difference between dipoles w/o modulation	−1.9984	1.9757	0.0202	0.4579
	Phase difference between dipoles with modulation	−1.9864	1.9926	−0.0088	0.9929
	Absolute phase difference between dip. w/o mod	0.0000	1.9984	0.2007	0.4121
	Absolute phase difference between dip. with mod	0.0000	1.9926	0.7940	0.5960
	Modulation effect	−1.9805	1.9030	−0.5933	0.7238
Corr.	R—Dipole inter correlation	−0.9982	0.9659	−0.0002	0.0728
	Absolute value of *R*	0.0000	0.9982	0.0156	0.0711
	Valid cases (list wise): 2880				

4.1 The Cluster Analysis

A cluster analysis of the observations was made as a function of the torque observed, the different modulating generators activated and modulation effect (Table 2).

Five different configurations have been extracted:

(1) Cluster SN at 4 Hz: characterized by the activation of both the 4 Hz generator and the 12 Hz generator
(2) Cluster Alfa On 12 Hz: characterized only by positive or negative activation of the generator at 12 Hz
(3) Composite DMN cluster: characterized by the activation of three generators: 2, 12 and 32 Hz
(4) Composite CEN Cluster: characterized by the activation of 4 generators: 8, 12 and 24 Hz
(5) Cluster Alfa Off 12 Hz: characterized by the absence of activation of the generator at 12 Hz.

The Cluster with the best performance under the experimental conditions as regards the modulation effect on the phase difference and the inter correlation effect on the amplitudes of the two dipoles appeared to be Cluster 3 DMN Composite, with generator activation at 2, 12 and 32 Hz.

The worst performance was instead observed for Cluster 1 SN at 4 Hz, with activation of the two generators at 4 and 12 Hz.

No significant effect was observed using the original modulation generators configurations.

Table 2 Modulation effect and inter dipole correlations by cluster

Cluster		*R*	Absolute (*R*)	Modulation effect
(1)	SN (Theta 4 Hz)	0.00070	0.00697	−0.49549
(2)	Alpha only	0.00056	0.00664	−0.52669
(3)	DMN (Delta2 + Beta16 and Beta32)	−0.00121	0.05275***	−0.76538***
(4)	CEN (Theta8 + Beta20 + Beta24)	−0.00095	0.01101	−0.68295
(5)	No alpha	−0.00051	0.01017	−0.52734
	Total	−0.00023	0.01561	−0.59331
	***$p < 0.001$	**n.s.**	**$p < 0.001$**	**$p < 0.001$**

4.2 The First Factorial Analysis: Modulation Activity

The association between activation of generators and the modulation effect or the correlation between Dipoles following modulation was studied, by factorial analysis (Table 3).

Five different factors were extracted:

(1) Configuration 1 at 8 and 20 Hz
(2) Configuration 2 at 2 and 12 Hz
(3) Configuration 3 at 4 Hz with harmonic inhibition at 24 Hz
(4) Configuration 4 at 32 Hz with harmonic inhibition at 16 Hz
(5) Configuration 5 at 4 Hz correlated to high Frequencies of both Dipole 1 and Dipole Two.

The activity of Configuration 1 and Configuration 2 appeared to be related to each other ($r = -0.094$; $p < 0.001$), in relation to the action of a common resonant frequency at 2 Hz.

The activity of Configuration 3 and 5 appeared instead correlated with each other ($r = 0.102$; $p < 0.001$) in relation to the common activity in the 4 Hz band.

These configurations appeared to express themselves in a significantly different way ($p < 0.001$) in the 5 clusters of activities, with methods consistent with the previous Cluster Analysis.

The Composite DMN Cluster 3 appears to be characterized by the activity of both Configuration 2 at 2 and 12 Hz and Configuration 4 at 32 Hz, while the CEN Composite Cluster 4 appeared characterized by configuration 1 at 8 and 20 Hz but by inhibiting configuration 3, i.e., activating the generator at 24 Hz and inhibiting the one at 4 Hz. The SN Cluster 1 at 4 Hz appears characterized by the inactivation of Configuration 2 (at 2 and 12 Hz) and by the activity of both Configuration 3 and Configuration 5, correlated to a high oscillation frequency of the two dipoles, with an activation of the generator at 4 Hz, but also an inhibition of the one at 24 Hz.

Table 3 Factors extracted in Analysis 1 by cluster

	FF1 (20/8 Hz)	FF2 (12/2 Hz)	FF3 (24/4 Hz)	FF4 (32/16 Hz)	FF5 (4 Hz)
Cluster 1 SN (4 Hz)	0.2091	−0.8280	1.1770	−0.0217	0.7832
Cluster 2 alpha (12 Hz) ON	−0.2525	−0.1413	−0.0967	−0.1632	−0.1720
Cluster 3 DMN (mixed)	−0.2321	1.4856	0.1539	0.6836	0.2529
Cluster 4 CEN (mixed)	0.6447	−0.1933	−0.6359	−0.1287	−0.1445
Cluster 5 alpha (12 Hz) OFF	−0.2605	−0.2084	−0.1061	−0.1507	−0.3257
	$p < 0.001$	$p < 0.001$	$p < 0.001$	$p < 0.001$	$p < 0.001$

The ME (negative value = best effect) appeared to correlate ($p < 0.001$) with the activity of Configuration 2 ($r = -0.081$), and Configuration 4 ($r = -0.130$), that modulating pattern at 2, 12 and 32 Hz: this is the characteristic activity observed in the Cluster DMN Composite, which in fact presents the best modulation behavior in terms of phase coherence.

The correlation of the activities between the two dipoles instead did not appear to correlate significantly with any of these modulation patterns, emerging only an inverse correlation ($r = -0.058$; $p = 0.002$) of the absolute IDC value for the factor characterized by high oscillation frequency and activation of the generator at 4 Hz, which is characteristic of SN Cluster 1.

4.3 The Second Factorial Analysis: Phase, Frequency, IDC and ME

Analyzing the correlation between the various evaluation parameters of the phase and frequency differences between the two dipoles and their relationship with ME and IDC, a five factors solution was found (Table 4).

(1) Phase difference without modulation (factor 1)
(2) Phase difference after modulation (factor 3)
(3) Oscillation frequency of the first dipole (factor 2)
(4) Second dipole correlation frequency (factor 4)
(5) Inter correlation between the activities of the two dipoles (factor 5).

ME appears to be in relation to the factor 1 ($r = -0.498$; $p < 0.001$) and factor 3 ($r = 0.913$; $p < 0.001$), expressing the action aimed at reducing the phase difference between the two dipoles, an effect that is most appreciated for the composite DMN Cluster 3 followed by the composite CEN Cluster 4. IDC instead appears

Table 4 Factors extracted in Analysis 2 by clusters

Factor	$\Delta\varphi$ w/o mod	Dipole 1 frequency	$\Delta\varphi$ with mod	Dipole 2 frequency	Inter dip. corr.
Cluster SN (4 Hz)	−0.3035831	0.0086766	0.0272336	0.2004086	−0.0504145
Cluster alpha (12 Hz) ON	−0.2473832	−0.0587385	−0.0007373	−0.0025745	0.0348191
Cluster DMN (mixed)	0.2791771	−0.0261780	−0.1484993	−0.0142086	0.0055961
Cluster CEN (mixed)	0.4729413	0.1015605	0.0432580	0.0015206	0.0205512
Cluster alpha (12 Hz) OFF	−0.1451024	−0.0038245	0.0560698	−0.1239644	−0.0449268
	$p < 0.001$	$p = 0.048$	$p = 0.010$	$p < 0.001$	n.s.

to correlate to the factor 3 ($r = 0.098$; $p < 0.001$) and factor 5 ($r = 0.781$; $p < 0.001$), which appears with no significant difference between dipoles. The phase difference following modulation (factor 3) appears to be the factor which interconnects the modulation effect with correlation between the two dipoles (factor 5), as the correlation between the two factor is significant ($r = -0.122$; $p < 0.001$).

4.4 The Third Factorial Analysis: Super Factors

Carrying out a study, by factorial analysis, of the interactions between the different modulation activity generators, the basic oscillation frequencies of the two dipoles and the observed effects on their phase difference, 4 different factors were extracted, in relation to the modulation action of 2 Hz generators either 4 Hz or 32 Hz (Table 5).

1. factor 1: integration of the two dipoles by 4 Hz generators as a function of the oscillation frequency of one of the two dipoles
2. factor 2: integration of the two dipoles by 4 Hz generators as a function of the oscillation frequency of the other dipole.

Table 5 Factors extracted in Analysis 3 (principal component analysis with varimax rotation)

	Integration by 4 Hz	Integration by 4 Hz	Integration by high beta when high $\Delta\varphi$	Dipole integration by 2 Hz
Explained variance (%)	*16.5*	*16.2*	*13.2*	*12.0*
Dipole 2 frequency	0.850			
FF3 (24/4 Hz)	0.496		−0.396	
Dipole 1 frequency		0.917		
FF5 ((f1 f2 4 Hz)	0.648	−0.721		
$\Delta\varphi$ w/o modulation			0.812	
FF4 (32/16 Hz)			0.596	
$\Delta\varphi$ with modulation				−0.618
R inter dipole corr.				0.600
FF2 (12/2 Hz)				0.477
FF1 (20/8 Hz)				−0.355

The two factors evaluated together indicate that the two dipoles can be integrated at 4 Hz in the range of frequencies used, if the frequency of one is low and that of the other is high.

3. The third factor expresses the integrability of the two dipoles in conditions of high phase difference, in the presence of 32 Hz modulation generators.
4. The fourth factor expresses the presence of both a high inter correlation of the activities between the two dipoles and a high modulation effect on the phase differences, in the presence of modulation generators at 2 Hz/12 Hz, but in the absence of activity of the generators at 8 and 20 Hz.

The frequency difference between dipoles is correlated with both factor 1 ($r = 0.571$; $p < 0.001$) and factor 2 ($r = 0.613$; $p < 0.001$). ME is correlated with both factor 3 ($r = -0.529$; $p < 0.001$) and factor 4 ($r = -0.544$; $p < 0.001$). IDC is correlated with factor 4 ($r = 0.352$; $p < 0.001$).

These 4 components are significantly different between clusters.

The composite DMN cluster 3, which appeared to have the greatest IDC and the greatest ME, appeared to have high values of the third and fourth factor, where the integration is obtained with oscillators at 2, 12 and 32 Hz (in the case of a high phase difference) while maintaining a good level of integration of the first dipole, thanks to its low frequency in relation to the range used (Table 6).

The CEN Cluster 4 has a high integrability component at 32 Hz, in the presence of a high phase difference between the two dipoles. The SN Cluster 1 has an integrative component at 4 Hz with a dipole oscillating at high frequencies compared to the experimental range, but at the same time, it does not have high phase difference integrability at 32 Hz.

Table 6 Factors extracted in Analysis 3 by clusters

	Integration by 4 Hz	Integration by 4 Hz	Integration by high beta when high $\Delta\varphi$	Dipole integration by 2 Hz
Cluster SN (4 Hz)	0.982	0.134	−0.643	−0.113
Cluster alpha (12 Hz) ON	−0.110	0.020	−0.237	0.006
Cluster DMN (mixed)	−0.065	−0.325	0.548	0.794
Cluster CEN (mixed)	−0.209	0.022	0.511	−0.460
Cluster alpha (12 Hz) OFF	−0.230	0.115	−0.185	−0.081
	$p < 0.001$	$p < 0.001$	$p < 0.001$	$p < 0.001$

4.5 Single Shot Simulations

By carrying out a Fourier analysis on the composite signal of the probe with about 80 Hz as average frequency of dipoles, following a modulation at 2, 12 and 32 Hz, a lowering of the final oscillation spectrum of the dipoles was observed as well as the appearance of a broad band of oscillation at 32 Hz. The change in phase difference (ME) was −0.19 to −0.93 ωt, while the maximum observed IDC was $r = 0.037$.

Increasing the phase shift between the dipoles (± 0.9 ωt), both the basic decrease in the oscillation band and the appearance of a band at 33 Hz are maintained, with an evident ME (from 0.11 to 0.02 ωt), while the maximum IDC was $r = 0.307$.

Increasing the oscillation frequency range of the dipoles increases the average frequency of their oscillation (90 Hz) and after modulation at 2, 4, 8, 20 and 32 Hz, the band at 32 Hz disappears, while two peaks at 50 and 60 Hz appear, with an increase of the maximum IDC to $r = 0.540$, in the absence of a significant ME.

Within a similar condition of mean oscillation (90 Hz) but with an increase of the phase shift between the dipoles (1 ± 0.1 ωt), and following a modulation at 2, 12 and 32 Hz, an inversion of ME (from 0.28 to −0.53 ωt) was found, while the maximum IDC increased to $r = 0.705$.

Finally, with settings more similar to the best pattern found previously (dipoles oscillating between 80 and 110 Hz, with a starting phase shift of 1 ± 0.1 ωt and a modulation pattern at 4, 20 and 32 Hz), the presence of modulation bands at 53, 65 and 96 Hz was found, as well as an evident ME (from −0.22 to 0.06 ωt) with a maximum IDC $r = 0.996$.

5 Conclusions

The final detected signal–very similar to the EEG pattern in the Gamma Band. This data is coherent with expectation since it is the result of the sum of the activities of Gamma Band Dipoles, modulated by low-frequency nodes. The model appeared also to be a useful prototype for studying theoretically the effects of low-frequency modulation activity on high-frequency dipole activity and how this effect favors a phase coherence and an activity inter correlation between the two.

It is clear from these first tests that the achievement of conditions of greater phase coherence and activity inter correlation may be linked to both the combination of the frequencies of the modulation generators and the oscillation frequencies of the dipoles, as well as on the degree of difference between both frequency and phase. These observations lead us to hypothesize on a theoretical level that Cross Frequency Modulation (CfM) activities can create conditions for integrating brain networks on the basis of preconditions that probably arise during development. This datum further suggests that with conditions of alteration of the CfM phenomena created during development following dysfunctions of the parental functionality, they can determine conditions of absence of the prerequisites for subsequent BNet full integration. This hypothesis is in line with the results reported by Marconi et al. (2018), in which differences in CfM between people with child maltreatment and normal controls were found. At the same time, these theoretical observations lead to the hypothesis that the cerebral electrical activity of each individual can contain the trace of his history. It highlights also that in the EEG spontaneous activity it can be detected the presence of system harmonic functional integrations or of an integration disorder which in turn may represent a psychopathological risk and the cause of decreasing of cognitive abilities. Further developments of this line of research imply further tests of the theoretical considerations presented here by analyzing the real EEG traces of high and low functioning subjects in terms of affective self-regulation and/or cognitive performance.

Further developments of this line of research imply.

1. The creation of more complex simulation models using programming methods more suitable for large complex calculations
2. Verification of the theoretical considerations presented by analyzing the real EEG tracings of high and low functioning subjects in terms of affective self-regulation and/or cognitive performance
3. Possible revision of the model on the basis of experimental evidence
4. Expansion of the model functions from simple network integration by phase coherence to possible information processing methods between input unit and output unit, verifying whether these integration methods really allow us to influence the transformation methods of the input signal into an integrated output signal.

References

Allen, E. A., Liu, J., Kiehl, K. A., Gelernter, J., Pearlson, G. D., Perrone-Bizzozero, N. I., & Calhoun, V. D. (2011). Components of cross-frequency modulation in health and disease. *Frontiers in Systems Neuroscience, 5*, 59. https://doi.org/10.3389/fnsys.2011.00059.

Axmacher, N., Henseler, M. M., Jensen, O., Weinreich, I., Elger, C. E., & Fell, J. (2010). Cross-frequency coupling supports multi-item working memory in the human hippocampus. *Proceedings of the National Academy of Sciences of the United States of America, 107*(7), 3228–33. https://doi.org/10.1073/pnas.0911531107.

Bressler, S. L., & Menon, V. (2010). Large-scale brain networks in cognition: Emerging methods and principles. *Trends in Cognitive Sciences, 14*(6), 277–290.

Hunt, M. J., Kopell, N. J., Traub, R. D., & Whittington, M. A. (2017). Aberrant network activity in schizophrenia. *Trends in Neurosciences, 40*(6), 371–382. https://doi.org/10.1016/j.tins.2017.04.003.

Klimesch, W., Schack, B., Schabus, M., Doppelmayr, M., Gruber, W., & Sauseng, P. (2004). Phase-locked alpha and theta oscillations generate the P1-N1 complex and are related to memory performance. *Brain Research. Cognitive Brain Research, 19*(3), 302–16. https://doi.org/10.1016/j.cogbrainres.2003.11.016.

Lisman, J. E., & Jensen, O. (2013). The θ-γ neural code. *Neuron, 77*(6), 1002–16. https://doi.org/10.1016/j.neuron.2013.03.007

Marconi, P. L., Pessa, E., Penna, M. P., & Tambelli, R. (2016). The phase amplitude coupling to assess brain network system integration (pp. 1–6). https://doi.org/10.1109/MeMeA.2016.7533783.

Marconi, P. L., Pessa, E., & Penna, M. P. (2018). Comparing different methods of measurement in psychiatry and clinical psychology: The diagnostic power of the new neurophysiology techniques. In *MeMeA 2018—2018 IEEE International Symposium on Medical Measurements and Applications, Proceedings*. Art. 8438667. https://doi.org/10.1109/MeMeA.2018.8438667

Melloni, L., Molina, C., Peña, M., Torres, D., Singer, W., & Rodriguez, E. (2007). Synchronization of neural activity across cortical areas correlates with conscious perception. *Journal of Neuroscience, 27*(11), 2858–2865.

Modolo, J., Hassan, M., Wendling, F., & Benquet, P. (2020). Decoding the circuitry of consciousness: From local microcircuits to brain-scale networks. *Network Neuroscience, 4*(2), 315–337. https://doi.org/10.1162/netn_a_00119

Moran, L. V., & Hong, L. E. (2011). High vs low frequency neural oscillations in Schizophrenia. *Schizophrenia Bulletin, 37*(4), 659–663.

Raichle, M. E. (2011). The restless brain. *Brain Connect, 1*(1), 3–12. https://doi.org/10.1089/brain.2011.0019. PMID: 22432951; PMCID: PMC3621343.

van Wijk, B. C., Jha, A., Penny, W., & Litvak, V. (2015). Parametric estimation of cross-frequency coupling. *Journal of Neuroscience Methods, 243*, 94–102. https://doi.org/10.1016/j.jneumeth.2015.01.032. Epub 2015 Feb 9. PMID: 25677405; PMCID: PMC4364621.

Systemic Cognition

Affective Matter: The Unconscious Between Brain and Mind

Vinicio Busacchi

Abstract The problem of affect and affectivity plays a significant role in psychoanalysis, both at a theoretical and therapeutic level. In Freud's first interpretation of hysterical symptoms affect is considered as a certain amount of energy, and this already mirrors its aporetic intertwining with the "reality" of the unconscious. For Paul Ricœur, Freud's psychoanalysis has a double, irreducible theoretical-procedural register, *energetistic* and *hermeneutical*; and his research into the unconscious reality reveals the significance and consequences of this double/dualistic approach. A combination of Husserl's phenomenological and Ricoeur's hermeneutical approaches helps us to find a "third, mediative way", in which the unconscious can be reinterpreted as a "affective matter", that is a mid-way reality placed *between* corporality and non-corporality, body and mind, natural causality and experiential/existential meaning. The notion of "affective matter" seems to indicate a sort of median, hybrid dimension of the unconscious as a "sphere of the unreflective" and as a "sphere of the instinctual" which is, a formula for rejecting Freud's *realism* of the unconscious without denying the instinctual and transformative *reality* underlying the life of consciousness. The problem to be solved is how the idea of transformation can be applied to the domain of identity transformation. I tackle transformation starting from its general reference to the process of the *symbolic appropriation* of reality.

1 Introduction

The problem of affect and affectivity plays a significant role in psychoanalysis, both at a theoretical and therapeutic level. In Freud's first interpretation of hysterical symptoms affect is considered as a certain amount of energy, and this already mirrors

---

V. Busacchi (✉)
Faculty of Humanities, University of Cagliari, Via Is Mirrionis 1, 09123 Cagliari, Italy
e-mail: busacchi@unica.it

G. Minati and M. Pietronilla Penna (eds.), *Multiple Systems*, Contributions to Management Science, https://doi.org/10.1007/978-3-031-44685-6_5

its aporetic intertwining with the 'reality' of the unconscious.[1] On the one side, it is true that, in his Metapsychology, Freud identifies the charge of affect with instinct and explains that only ideational representation, not the affection, is found at the unconscious level. This means that affectivity is something concerning the preconscious and conscious psychic life. However, on the other side, the existence of a 'representative' dimension within the unconscious opens the way of a dual or bifacial re-reading of it, beyond the biological-organic constitution which Freud assigned to the *Unbewusste*. (As is known, Freud does not take a clear-cut position in this regard: his project of a unilaterally neurobiological definition of the unconscious decays with the abandonment of his youthful project of a psychology for neurologists [1895]. Yet, it remains true that throughout his work, he thinks of the psyche in terms of a psychobiological reality [see, for, e.g. Sulloway, 1983]. This finds confirmation in Freud's final model of mental apparatus, where unconscious [i.e. the *Es*] does not refer only to the *quality* of a certain mental state nor does it have only the meaning of a *system*, nor is it valid only in a *dynamic* sense for the "struggle of forces" between contrary contents: the *Es* is a reality; it is the "reservoir" both for instinctual drives and desires and for [dynamically] repressed contents [see Sandler, Holder et. al. 1997]).

For Paul Ricoeur, Freud's psychoanalysis has a double, irreducible theoretical-procedural register, *energetistic* and *hermeneutical* (Ricoeur, 1970); and his research about the unconscious reality reveals the significance and the consequences of this double/dualistic approach. A combination of Husserl's phenomenological and Ricoeur's hermeneutical approaches helps us to find a "third, mediative way", in which the unconscious can be reinterpreted as an "affective matter", that is a mid-way reality placed *between* corporality and non-corporality, body and mind, natural causality and experiential/existential meaning.

This path does not pave the way for a new phenomenological-existential approach to the unconscious but rather to a multidisciplinary approach to the complexity of the unconscious as a *reality* operating both on a psychic level and on a neurobiological level, and both on an intrapsychic level and on a relational, intersubjective level. This multidisciplinary approach can represent the most concrete and exact response to the complexity of the human being's psycho-organic life, characterised by complex and varied processes. As Wilma Bucci points out among others, human beings possess multiple forms of information processing (Bucci, 2009: 29). These forms are not perfectly integrated as they configure "different states" and "different systems" variously implicated both in the assumption and processing of sensory and bodily information as in the processing and expression of inner, psychic and sensory-motor contents.

[1] As Laplanche and Pontalis explain, the notion of affect is recognised as having a central role in Breuer and Freud's essay on hysteria (*Studien über Hysterie* 1895), where the origin of hysteric symptom is identified with a blocked affection, that is an affective excitation connected to a meaningful event, which did not find an adequate discharge (see Laplanche, Pontalis 1967: t. 1, v. affect).

2 An Approach to the Unconscious

A theoretical approach of a phenomenological-hermeneutic type can reject both psychologistic and neuroscientist reductionism. There is a concrete possibility of conceiving the unconscious in its complexity through a plural approach to its reality. This is possible starting from the contemporary developments of research in the neuroscientific field. As Vittorio Gallese observes, "psychoanalysis has always identified the body as the main source for feeding psychic representations. Some recent developments in cognitive neuroscience research have [...] highlighted the importance of the body in action and of the sensory-motor systems in the constitution of the representation of reality" (Gallese, 2009: 198; en. tr. V.B.).

The most recent neuroscientific research restores significance to Freud's youthful project of a *Entwurf einer Psychologie für den Neurologen*, that is the idea of a formation of the unconscious at the level of memory. In particular, research in the field of infant research and around the idea of implicit memory confirms the "structuring force" of psychic experiences which (1) are "affectively charged" and (2) mainly take place in the period of psychic development. This research also confirms the correctness of Freud's youthful intuition, according to which nothing recorded in the mind is destroyed, rather it remains as a memory trace, at the level of what today is called unconscious memory (see Moccia 2009: 115 f). This deep level of brain and memory functioning also includes the development and structuring of mutual affective modulation skills of primary bonds. As Solano points out, "the idea is in the psychoanalytical tradition (Winnicott, Bion, Tustin, Gaddini) that the ability to contain, modulate, think and process affects is closely connected to the experiences of understanding, sharing and mirroring affective states by primary objects in the early stages of life. [...] The child comes to contain and regulate his negative affects through the internalisation of this transformative function of the mother" (Solano, 2009: 115,116; en. tr. V.B.). This process of internalisation can certainly be placed in parallel with the idea of habit according to the meaning identified by Bateson (Bateson, 1967). This meaning opens the way to an understanding of the unconscious that is also valid under a systemic approach (see Casadio, 2018) and, therefore, cannot be reduced to the psychological or psychobiological meaning alone.

Of course, the notion of "trace" remains highly metaphorical. Among the various scholars, Ricoeur has focused on the topic, showing in particular the coexistence of a material dimension and a factual/experiential dimension in the semantics of the trace (see Ricoeur, 2004). The notion of trace, in fact, applies both to *mnestic trace* and to *historical trace*, i.e. both as a biochemical mark and as an experiential mark [= sign of an experience, remembrance]. This distinction goes back to the difference between Aristotle and Plato in understanding memory as a *remembrance* rather than an *impression* ('impression' like from a seal on wax). Ricoeur connects this discourse to the problem of personal identity, which, far from being able to be defined in purely substantial or purely psychological terms, sees the link as close (1) between the *natural* level of identity and the historical-experiential level of the *person*, and (2) between the subjective plane and the relational plane (see Busacchi,

Martini 2020). It is no coincidence that Ricoeur has intertwined his theory of the "capable human being" (*homme capable*) with a reinterpretation of psychic life according to Freud's psychoanalysis through a key that accentuates the dimension of the relationship and of recognition (Ricoeur 2005). He does so first of all by deepening Heinz Kohut's research (Kohut, 1984) on *consciousness*, *ego* and *self*, and the attention he dedicates to the relationship with the psychic and to the structuring capacity of the intersubjective relationship (see Ricoeur, 1986).

3 Towards a New Metapsychology?

The notion of trace remains as metaphorical but asserts itself in its effective capacity to reflect something intrinsically dual, something that cannot be conceptually determined in a unilateral way (under penalty of a reductionist positioning). On this dual dimension, which cannot be reduced to the biological or the psychic alone, the understanding of the unconscious in its essentiality is maintained. This is how the possibility of a new, *psychobiological* synthesis approach to the unconscious opens up. This is what, e.g. the American psychoanalyst Wilma Bucci proposes. Bucci's work is an attempt to profile a renewed Metapsychology *through* contributions from cognitive science and neurobiology. Bucci develops an *integrated* model based on a conception of psychic functioning centred on the category of multiplicity, with reference to a new psychobiological order to which the sphere of the so-called *sub-symbolic* pertains (Bucci, 2009). The symbolic is further articulated into verbal symbolic, i.e. the productions concerning the sphere of speech and language, and non-verbal symbolic, in which imaginative activity finds expression at the level of sensoriality (mainly visual). This is the field of elaboration and re-elaboration, representation and manipulation of meaning according to the different expressive-representational modalities of sensorial-psychic life. The sub-symbolic system, connected to the previous one, already manifests a function of processing and re-processing of information (Bucci, 2009: 30). What is of interest in psychoanalysis for Bucci is the centrality of sub-symbolic elaboration in the processing of emotional information and in emotional communication (Bucci, 2009: 31). It is an aspect (by quoting Daniel Stern) that she develops with reference to infant research, and which again brings us back to the relational dimension and to the "dialectic" between verbal/non-verbal and the affect. The phenomenon of "affective attunement" described by Stern is essentially a type of continuous and analogical emotional communication (*ibidem*), i.e. according to the modality of sub-symbolic representation/expression. Bucci underlines the distinction between sub-symbolic elaboration processes and between primary process: unlike the primary process studied in psychoanalysis, sub-symbolic processes are not chaotic, they are not oriented towards the satisfaction of desire and they are not segregated from reality. Sub-symbolic processing can take place both inside and outside awareness (*ibidem*). However, it is clear that her model goes beyond the level of psychodynamic discourse. The symbolic and sub-symbolic life are connected to and rooted in sensorimotor life. This means that the (pre-)representational cannot be explained only in terms of the consciousness-unconsciousness dynamic and that the very conception of psychic reality needs to be reconsidered. In accordance with Bucci's perspective, a theoretical itinerary based on the phenomenological-hermeneutical approach leads its heuristic effort into a new understanding of psychic reality in the new terms of "affective matter". It is precisely the idea of affective matter that proves to be strategic and fertile for an approach capable of reflecting the complexity of psychic life.

4 Affective Matter

Ricoeur's youthful research on the *voluntary* and the *involuntary* is strategic in highlighting that there must be a link of continuity and non-duality between the dimension of intentionality and will and between the dimension of the unconscious understood in phenomenological terms as a sphere of pre-intentionality, that is as a sphere characterised by a «matière principalement affective» (Ricoeur, 1966). This idea comes from Edmund Husserl who combines Freud's unconscious with the Greek notion of *hyle*, a material "affectively charged", i.e. the sphere/dimension of the vectorial and the pre-intentional (*Ideen II*, 1952 [post.]).

The notion of "affective matter" seems to indicate a sort of median, hybrid dimension of the unconscious as a "sphere of the unreflective" and as a "sphere of the instinctual"; that is, a formula for rejecting Freud's *radical realism* of the unconscious without denying the instinctual and transformative *reality* underlying the life of consciousness. This idea is also suggested by the reference to the Greek term. It is known that both in Plato and Aristotle, the concept of "matter" oscillates between the idea of a given reality, or of matter-*subject*, and the idea of receptivity, or of matter-*power*.

The problem to be solved is how the idea of transformation can be applied to the domain of identity transformation. So far, I tackle transformation starting from its general reference to the process of the *symbolic appropriation* of reality (= the painter who paints a landscape). Then I address, following Bion, its application to the psychoanalytic process. I then arrive, referring to Ricoeur, at the idea of an *affectivité puissante* that presses on the boundary between soma and psyche and is constituted, like Freud's *Rapräsentanz*, as a link between the unreflective and the intentional. Finally, moving to the level of neuropsychology, I explore the transformative function exercised by the cerebral cortex on perceptions and the unconscious to bring them to consciousness.

This interpretation fits into the general design of a theoretical-speculative articulation between the mental/inner world governed by the law of motivation and sense and the psycho-physical world governed by the causal law and by the biological mechanism. The unconscious is understood as connected both to psycho-(pre)intentional aspects and to psycho-physical aspects, so that with the "pre-intentional" both the drive and the sphere of instinctual desires are reinterpreted. Such an interpretative line also modifies the understanding of the notion of representation, which is no longer explainable only in terms of configuration and hermeneutics. The representation emerges as a sort of conveying process both of *affectivity* and *sense* as (following Bucci) of sub-symbolic elements, that is sensory-cognitive-motor.

5 On the *Vorstellungsrepräsentanz*

For authors such as Ricoeur, the nexus of ideational representation constitutes the theoretical crux of the problem of the ontological (and epistemological) foundation of Freud's psychoanalysis. Ricoeur conceives the concept of *Repräsentanz* as an expression of the "coincidence" between the dimension of meaning and the (organic-biological) dimension of force. It is this coincidence that explains the finality or "destiny" of the drive in the psychic representation. Drive and instinctive desire arrange/orientate themselves towards the "representational expression". It is, evidently, a passage that is a *transformational process*. As Ricoeur explains,

Freud's originality consists in shifting the point of coincidence of meaning and force back to the unconscious itself. He presupposes this coincidence as making possible all the "transformations" and "translations" of the unconscious into the conscious. In spite of the barrier that separates the systems, they must be assumed to have a common structure whereby the conscious and the unconscious are equally psychical. That common structure is precisely the function of *Repräsentanz* (Freud 1970: 135).

It is true that Ricoeur thinks that the functioning of the *Repräsentanz* is possible by virtue of a substantial uniformity between the conscious structure and the unconscious structure. The concept of *Repräsentanz* would indicate the "place" where the instinct designates itself: it is understood as something "psychic" that represents the "instinct" as energy; more precisely, the bridge-concept is that of *Vorstellungsrepräsentanz*—translated by Ricoeur with *présentation représentative*. With this, Freud would show that he has reintegrated the unconscious into the sphere of meaning. However this line of reasoning—which produces a certain persuasive effect in Ricoeur's essay—fails on the key node of deepening the meaning and use of the concept of *Vorstellungsrepräsentanz* in a context, that of Freud's Metapsychology, which closely links psychology and biologism. Limiting oneself to translating as "representative presentation" (or "ideational representation") and giving it a sense of substantial linguistic-symbolic character and intrinsic significance, without trying to "meet" Freud's theoretical-scientific point of view raises some doubts about the operation field. This happens even because the concept of *Vorstellungsrepräsentanz* shares a certain familiarity with the notion of *psychische Repräsentanz*, "psychic representation", and *Triebrepräsentanz*, "representation of the drive". The effort made by Laplanche and Pontalis in the *Vocabulaire de la psychanalyse* appears more valuable. In it they propose an understanding of the "ideational representation" as a delegating-delegated relationship, i.e. as a dialectical-relational nexus where a certain "thrust" causes, determines and expresses something representational. For Laplanche and Pontalis, the relationship between somatic and psychic is neither conceived on the model of parallelism nor on that of a causality; it must be understood on the basis of the comparison with the relationship existing between a delegate and his principal (see Laplanche, Pontalis 1967: v. représentation psychique). The two authors put forward the hypothesis that the difference which can be seen between them is only verbal: in the first case the psychic representation would be called ideational representation, in the second case it would be called drive.

With this last statement, Laplanche and Pontalis convey a certain interpretative commitment, careful not to lose the substance of the energetic discourse and the reference to the dimension of the drive and the body. Yet the question of psychic representation can be reinterpreted within a framework that takes into account both the discourse of the non-duality of the mental and the body (in the sense of their mutual connection) as well as the complexity of the systems characterising the functioning of the human organism and psyche. Ricoeur's thesis according to which the functioning of *Repräsentanz* is possible only on the postulate of a structural uniformity between unconscious and conscious appears weak. On the contrary, it has been highlighted from many sides how Freud himself spoke of a constitution in qualitatively different layers of the apparatus (Sandler et al., 1997). In addition, Freud's vision also has to be rethought both by leveraging the re-reading that offers a phenomenological-hermeneutic approach (= unconscious as affective matter) and by taking into account studies such as that of Bucci. These studies reveal that the complexity of the functioning of the human organism derives from an intersection, superimposition and connection of states and systems such that it is not possible to think of the symbolic as a *purely* linguistic-representational instance, of the representational as a proper dimension of only mental life, to the affective as something predetermined and not culturally (re-)mouldable, etc.

6 In Conclusion

Proceeding from this reinterpretation of the unconscious that takes into account the complexity and non-duality of the organic-psychic life, the possibility emerges of understanding the relationship between unconscious and the conscious as mutually influenced according to a transformational process. Between the contents of the biological sphere and the contents of the psychic sphere, between the cerebral reality and the psychic reality, between the natural dimension and the dimension of meaning, there is not only a relationship of profound connection and coordination but a profound procedural "dialectic" on the mediation of the affective and the representational. If this is the case, neurobiological functioning is just as central to the entire psychic life as psychic life is central to neurobiological functioning. It is in particular the deepening of the reality of reinforced concrete and of the entity of psychic representations that paves the way for a non-dualistic and corresponding-transformational model in the understanding of human psychic life and the relationship between the mental and the psychical.

References

Bateson, G. (1967). Style, Grace, and Information in Primitive Art. In: G. Bateson (Ed.), *Steps to an Ecology of Mind.* New York: Ballantines Books 1972.

Bucci, W. (2009). Lo spettro dei processi dissociativi. Implicazioni per la relazione terapeutica. In: G. Moccia, Solano, L. (Eds,), *Psicoanalisi e neuroscienza. Risonanze interdisciplinari.* Milano: FrancoAngeli, 29–53.

Busacchi, V., Martini, G. (2020). *L'identità in questione.* Milano: Jaca Book.

Casadio, L. (2018). L'inconscio sistemico. In: *Connessioni*, 4, December 2018; http://connessioni.cmtf.it/linconscio-sistemico/ (last accessed: 2/2/2023).

Gallese, V. (2009). Simulazione incarnata, intersoggettività e linguaggio. In: G. Moccia, L. Solano (Eds.), *Psicoanalisi e neuroscienza*, cit., 174–206.

Kohut, H. (1984). *How does Analysis Cure?*, ed. by A. Goldberg and P. Stepansky, Chicago: University of Chicago Press 1984.

Laplanche, J., & Pontalis, J.-B. (1967). *Vocabulaire de la psychanalyse.* Presses Universitaires de France.

Ogden, T. H. (1994). *Subjects of Analysis.* Karnac Books.

Ogden, T. H. (1997). Some Thoughts on the Use of Language in Psychoanalysis. *Psychoanalytic Dialogues, 7*(1), 1–21.

Ricoeur, P. (1966). *Freedom and Nature: The Voluntary and the Involuntary.* Trans. E. V. Kohák. Evanston (IL): Northwestern University Press.

Ricoeur, P. (1970). *Freud and Philosophy: An Essay on Interpretation.* Trans. by D. Savage. New Haven and London: Yale University Press.

Ricoeur, P. (1986). The Self in Psychoanalysis and in Phenomenological Philosophy, *Psychoanalytic Inquiry*, n. 3, pp. 437–458; now in Ricoeur, P., *On Psychoanalysis*, trans. by D. Pellauer, Cambridge: Polity Press 2012, pp. 73–93.

Ricoeur, P. (2004). *Memory, History, Forgetting.* Trans. by K. Blamey and D. Pellauer. Chicago: The University of Chicago Press.

Ricœur, P. (2005). *The Course of Recognition*, trans. by D. Pellauer, Cambridge (Mass.): Harvard University Press.

Sandler, J., Holder, A., Dare, C., & Dreher, A. U. (1997). *Freud's Models of the mind.* International, Universities Press.

Solano, L. (2009). Teorie psicoanalitiche del funzionamento mentale inconscio e teoria del codice multiplo, in G. Moccia, L. Solano, *Psicoanalisi e neuroscienza*, cit.: 113–124.

Sulloway, F. (1983). *Freud, Biologist of the Mind.* Basic Books.

Intelligent Systems Many Manners of Adapting to Environment

Lucia Urbani Ulivi and Primavera Fisogni

Abstract What is intelligence is a debated and still open issue. Psychologists have identified many different skills involved in intelligence, but there is no agreement on a general definition. Without entering in the debate, we will here adopt a definition pertaining to philosophical area, general enough to include most meanings currently used in psychology and in common sense and specific enough to distinguish intelligence from other capacities. *Intelligence is the ability to find adaptative responses to perturbations both of the internal and the external environment recovering proper equilibrium.* Most systems, artificial and natural, are intelligent in the sense of the proposed definition, as most of them have the capacity to restore their systemic balance in response to change occurring in both inner and outer milieu. We shall then investigate by which means different systems adapt to their environment and if—and in what aspects—human intelligence differs from every other kind of intelligence known to us. To answer both questions, we will distinguish objective—or intrinsic, or tacit—intelligence from subjective—or explicit—intelligence, characterized by mental activity, language, and auto conscience. We will finally support the thesis that in human being intelligence emerges thanks to the interplay among three actors: objective intelligence, subjective intelligence, and their environment(s).

Keywords Open systems · Intelligence · Objective intelligence · Subjective intelligence · Emergence

L. U. Ulivi (✉)
FTL-USI Lugano, Lugano, Switzerland
e-mail: urbaniul@teologialugano.ch

P. Fisogni
La Provincia di Como, Como, Italy
e-mail: pfisogni@gmail.com

G. Minati and M. Pietronilla Penna (eds.), *Multiple Systems*, Contributions to Management Science, https://doi.org/10.1007/978-3-031-44685-6_6

1 Introduction

What is intelligence is a debated and still open issue. Psychologists have identified many different skills involved in intelligence, but there is no agreement on a general definition. In searching for a collection of definitions, Legg and Hutter (2007) found 70 different ways in which intelligence can be said, spanning from an individual property that interacts with the environment to the ability to succeed with respect to some goals.

Without entering in the debate, we will here adopt a definition pertaining philosophy, general enough to include most meanings currently used both in psychology and in ordinary language, and specific enough to distinguish intelligence from other capacities.

> *Intelligence is ability to find adaptative responses to perturbations both of the internal and the external environment recovering proper equilibrium.*

Most systems, artificial and natural, are intelligent in the sense of the proposed definition, as most of them have the capacity to restore their systemic balance in response to a change occurring in both the inner and the outer milieu, but each system adopts different strategies to catch such goal.

Within the systemic realm a further step is to frame this investigation into the multiple systems domain (Minati & Pessa, 2006), in which the elements that form the system itself play interchangeable roles and interact simultaneously or sequentially in different ways, giving rise to sequences of different systemic conducts and to different systems. Multiple systems are called collective beings when the elements are autonomous, being able to decide how to interact. An example that is in the experience of all of us is given by the Internet (Minati, 2022: 110).

This is not simply a kind of methodological premise. On the contrary, the dynamic, adaptable, and interchanging notion of intelligence sketched before should be thought as a system, or it would be better said, in terms of *multiple intelligent systems* with a plural identity, albeit grounded on a set of common features.

From this perspective, the present investigation is expected to introduce some remarkable novelties for both philosophy and systemic thinking that might bring about a change in the debate. Precisely, philosophy is invited to move forward the comfort zone of the cognitive approach to intelligence, focusing on the multiple layers of such a dynamic ability to reach a comprehensive view on animated/inanimate beings.

On the other hand, in systemic thinking it can be introduced the concept of *intelligence of the emergences* as both (1) the engine that identifies the establishment of coherence and self-organization in collective systems and (2) the process that generates singularities. To postulate such an intelligent texture, beyond giving reason of a 'greater systemic integration and harmonization' (Marconi & Penna, 2021: 87), can be valuable for explaining the failure of predictive conceptual frame only recurring to linear paradigms. An example? The Black Swan phenomena (the Covid-19 outbreak or the 9/11 just to quote two of them), whose dynamics seem to escape linear thinking (cause/effect) and belong to an implicit, however, well-directed strategy.

2 Intelligence is a General Feature of Open Systems

The first step to move into the multifaceted profile of intelligence calls for a short account of open systems: they radically differ from formal systems, which are closed, since the rules that establish them determine their outcomes in a complete and predictable way.

On the contrary, the interaction with the environment through processes of dissipation and acquisition is the landmark of open systems. Multiple filters of various levels allow open systems to maintain exchange relationships with the environment and to valuably support the acquisition of emerging systemic properties. All of them are, for the observer, the cues that the system is intact despite its process variations.

The capacity to be transformed through the interaction with the world, and at the same time to sustain its own integrity, makes of an open system a highly adaptive system. This complex ability in which not only homogeneous, but also heterogeneous aspects are finely tuned, should be seen as the most characteristic feature of what is generally viewed as intelligence.

Nevertheless, intelligence may be specified in different ways according to the type of system described. Intrinsically related to it, the creativity of emergences deserves a brief focus. It presents itself as unexpected, unpredictable, surprising, through the extraordinary variety and richness of the adaptation processes deployed by natural entities and to a certain extent also by artificial ones. We can conclude 'intelligence', 'adaptivity', 'creativity' are terms that refer to the same property, while each point out a different aspect of such ability.

Worth noting that with 'adaptivity' we refer to the ability to choose between different possibilities of answering a personal need or a question posed by the internal or external environment in the form of perturbation, while the thick term 'creativity' underlines the originality of the adaptive response, which invents forms, objects, scenarios, projects—and much more—entering the processes of the world and transforming them. 'Intelligence' especially refers to the capability to connect untied elements giving birth to a new scenery. Intelligence, adaptivity, and creativity undermine the possibility of understanding the behavior of open systems in a deterministic way and open to the vast and articulated spaces of freedom.

Intelligence—or adaptiveness, or creativity—is a common feature of the open systems that exhibit emergences. Through the lenses of systemics we can easily realize that there is no system without emergence; the ability to withstand emergences shows that a system to keep itself intact is using intelligence, adaptability, and creativity. Being a system, then, implies being intelligent (whether the converse is also true is a question that should be investigated).

Although intelligence is a common feature of open systems, it is far from being identical in all of them; on the contrary, each system uses such capability in its own manner, opening new paths and finding specific solutions to the ever-repeated problem of successfully kip connections with its environment. The way a giraffe is intelligent highly differs from the way a human is, and the degrees of adaptation are very different from one human to another. Given that intelligence is multiple, we want now to investigate in what way does human intelligence dwell a specific place in the rich scenery of intelligences active in our world.

3 Objective—or Intrinsic—Intelligence. The Tacit Dimension of Intelligence

When we say that we as humans are intelligent, we typically refer to the intelligence we are directly acquainted with, observable in what we do, in what we say and in what we think. In the terms introduced by Polanyi (1966) we are pointing to explicit intelligence, the one that has been exclusively or primarily taken in consideration by philosophy and entered the cultural discourse as a property that can be grasped and described discursively. But beside it, a tacit intelligence must be staged if we want to track the roots that nourish our explicit intelligence and if we want to explain the successful behavior of a system—both human and not human—in the reference environment. The need to consider an intrinsic intelligence has been reduce the space between the two lines neglected for centuries, with the remarkable exception of Aristotle, who distinguishes *noesis* from *dianoia*.[1]

If Freud was not the first to suggest the existence of a submerged ground of reality, not detectable in human self-observation, he was certainly the first to forge some useful tools for becoming aware of that realm. Nevertheless, the Freudian submerged dimension of reality mainly pertains to the sphere of emotional dynamics: it will be necessary to wait many decades before the neuroscientist Libet (2004) was in the condition to formulate the hypothesis of an unconscious mental activity and to test it experimentally.

Libet observed that many phenomena well known to scientists let us glimpse the tacit intelligence at work, as among others the correction of neuronal distortions, the solution of problems without the activation of conscience, the memory of something that was said when the subject was under anesthesia. It is also active in blind perception—when the body is oriented to reach the target without his movement being guided by conscious processes. These, and other similar behaviors, are performed according to a rational structure able to act pursuing a goal and cannot be understood in reductive terms such as chaotic or even occasional or simply casual phenomena.

To better grasp intrinsic intelligence, we shall start a trip, leaving the explicit dimension of the real to dig into the vast and uncertain domain of the implicit, guided by the Ariadne's thread of reason.

No doubt that the pandemic outbreak made the *implicit* or *objective* or *tacit intelligence* 'viral' and more actual than ever, even if such presence has been scarcely recognized. A radical shift occurred within the immunization processes in order

[1] *Noesis* is the basic level of knowledge, consisting in the immediate apprehension of things. This level ensures direct contact with the domain of ideas and of reality. The second level, called *dianoia*, is the realm of judgment, where the unity of *noesis* is splitted to form propositions and logical inferences.

to struggle the virus: two antiCovid-19 vaccines (Comirnaty by Pfizer; Spikevax by Moderna) used messenger ribonucleic acid (mRNA) molecules that contain instructions addressed to the cells in order to synthesize Spike proteins.

According to this microbiological path, it is not the virus to be introduced into the cell of the person, but the genetic information that the cell needs to build copies of the Spike protein. The method is as simple as effective. Blocking this protein, with which SARS-CoV-2 typically attacks the immune system, it can be reached a main goal. The immune system can recognize the virus, stopping it or at least reducing its powerful impact on the organism.

Objective intelligence is at work to reach such complex goals, since all these operations are not mechanical; mRNA vaccines need a certain capacity to be acquainted with the new environment, as well as the ability to spread information in time and to know the way to reach the right targets.

Many other phenomena can be better explained if we follow the path of intrinsic intelligence and could be an exciting exercise of investigation to detect other mystery cases losing their opaqueness if we shift from the paradigm 'intelligence is present to conscience' to the more comprehensive claim 'intelligence is at work also when nobody is aware of it'.

By the way of examples, we shall briefly investigate three complex systems, belonging to the ecosystem and the hyperconnected world, to detect how objective/implicit intelligence works, according to what relevant processes, with what results. Such systems are phototropism, metabolism, and cloud computing. Phototropism and metabolism play crucial roles in the world of life, while cloud computing dwells in the *onlife* domain, (Floridi, 2015) where the real and the digital are melted together.

3.1 Phototropism and the Intentional Responses of Plants

Phototropism (light + movement), the way green plants respond to the light source, is a biological phenomenon that can be easily observable. Let us think of a seedling kept in a room with little lighting: its stems are oriented toward the most intense light stimulus; if we change the direction of the light beam, the plant is able to perform a turn in that direction. It seems to move in a deliberate way, to ensure flowers, leaves, trunk, roots the optimal growth conditions.

Clearly revealing of the adaptive behavior within the botanic realm, phototropism has become a relevant topic for a better understanding of the intelligent behaviors of green plants, rising several issues in the direction of plants cognition. The lack of a neuronal system, in fact, is no more considered sufficient reason to exclude a priori these living entities from the highest level of capabilities usually only attributed to the more evolved animated beings.

Among scholars there is a wide consensus that the photosynthetic and the phototropic world of plants is related to specific organs that collect, organize, and remodulate the external light stimuli, giving rise to the ability to choose among different and conflicting options, as well as the animated beings do. On one side, the focus is on the cells as the barycenter of the neuronal-like skills (Trewavas, 2009, 2014, 2016); on the other side, a 'root-brain' is seen as the central infrastructure of the plants (Baluška et al., 2008). They both search for a special type of cognition in plants.

According to the first theory: «intelligent behaviour exhibited by single cells and systems similarity between the interactome and connectome indicates neural systems are not necessary for intelligent capabilities». Intelligent leaf movement in low light, adaptively plastic and intelligent responses to light, intelligent responses to light resources, and adaptive variability to light in response to crowds can be experimentally detected. In these investigations, rooted in Darwin's intuitions, (1880) resound the discourse of plant biologist McClintock, on the occasion she was awarded the Nobel Prize.

> «A goal for the future would be to determine the extent of knowledge the cell has of itself and how it uses that knowledge in a thoughtful manner when changed» (McClintock, 1984, quoted from Trewavas, 2016: 2).

Plants neurophysiology, the other scientific path into plants intelligence, is grounded on plant neurobiology, in which many neuronal-like activities, based on plant synapses, appear to be not only theoretically justified, but experimented (Baluška et al., 2006; Baluška & Mancuso, 2007). A network of synapses, for example, transport auxin, the hormone for the growth related to phototropism. Roots' apices «emerge as command center» while «act as the posterior poles». In conclusion:

> «As plants are capable of learning and they take decisions about their future activities according to the actual environmental conditions, it is obvious that they possess a complex apparatus for the storage and processing of information» (Baluška et al., 2006: 19).

Beyond the different paradigms—cells or root-brain oriented—both the perspectives recognize in the light-foraging behaviors, a proper way of learning by association (Gagliano et al., 2016), an *intentional* response of plants, that is to say a finely grained skill of the intelligent beings. Intention, in fact, is the kind of knowledge that allows to reach a goal. For Anscombe (1957) intention is an act belonging to the class of «non-observational knowledge», or the «class of movements known without observation» (Proposition 8). Just to sketch an example, when we say «I open the windows» I make several operations in order to perform that act. What does it mean, from a cognitive perspective? Simply that intention is grounded on knowledge of the means with which a goal is achieved (to open a window) (Torralba & Lano, 2008, 2010). However, one thing is to say that plants are not passive systems, one is to relate their somehow sophisticated forms of behavior to a specific agency, goal-directed (intention), and another one is to follow McClintock when she suggests that plant cells not only have knowledge, but can know themselves. The notion of intrinsec intelligence can usefully support the first two statements, while McClintock hypothesis that plant cells can think themselves seems derived from a monistic concept of intelligence that should be abandoned in favor of a pluralistic idea of intelligence.

Actually phototropism—as the result of an effective, intentionally directed interaction of multiple systems (plant, environment, light, and metabolism)—recalls cognitive processes that are performed without a clear consciousness: through appetite, for example. In classical philosophy, this idea has been theorized by Thomas Aquinas in the doctrine of *connaturalitas*, which refers to the judgment *per modum inclinationis* and differs from the intellectual judgment, aka *per modum cognitionis* (Biffi, 1974). Natural inclination is recognized to be present everywhere in nature (Baldner, 2018) and strictly related to appetite, exactly the kind of 'movement' related to plant tropism toward a source of light (Keane, 1966).

3.2 Metabolism: Informational Exchange with the Environment

Metabolism could be understood as the most powerful engine for life growth because its processes are oriented to support living organism through biological reactions. At large, metabolic functions are also intended in a metaphorical way, as in the case of urban metabolism, that concerns flows within cities, in order «to quantify he inflows, outflows, and accumulation of resources (such as materials and energy) in a city» (Derrible et al., 2021: 85). The phenomenon of phototropism sketched above is a typical metabolic response of green plants to solar energy.

Biological reactions are hosted by each cell through a set of operations spanning from the transformation of food into energy, through enzyme-mediated responses, to the elimination of wastes. In biological terms, these activities depend upon two related phases, the catabolic reactions (the breakdown of bigger molecules into smaller) and the anabolic processes (the combination of simple molecules into bigger ones).

Among the multiple processes that characterize metabolism, the leading one is about the capacity to synthesize new materials/molecules, precisely amino acids, carbohydrates, and nucleic acids (the components of the DNA) and the proteins. This basic set of information throws light on metabolism as an 'agent' capable of a special type of intelligence (someone would prefer to speak of 'cognitive abilities', or of 'cognition'), selecting what is good and what is bad for life to flourish.

It is simply impressive to notice the way biological processes enzyme-mediated transform their own environment. The notion of complex system perfectly fits to metabolism, a process that can only take place in the macro-system of life that influences so many subsystems and acts through structural dynamics, showing itself in terms of a sequences of phase transitions. At least four phases can be detected:

(1) **Change of structure** It occurs, for instance, in the passage from food to new molecules.
(2) **Acquisition of a structure**, or passage from an unstructured to a structured configuration. In phototropism, the solar energy allows the production of auxine, the hormone that causes the cells of the plant to elongate on the shaded side. In this case, the acquisition concerns both a new material (the hormone) and a new organic process (the growth).

(3) **Loss of structure**, as for instance transition from a structured to an unstructured configuration. Pathogenesis of neurodegenerative disorders, as a growing body of evidence seems to prove, depends on energy metabolism decline in aging (Błaszczyk, 2020).
(4) **Combination of structures**. They continually occur in metabolism. In human metabolism, the main business is given by the oxidation process. In the last phase, the transformation of indigestible dietary residues occurs through fermentation, and the leading actor becomes the intestine itself. This example is particularly interesting for diving into the hidden life of metabolic intelligence, not simply because it presents the search for an alternative way to reach a main goal (fermentation instead of oxidation), but because it unveils a functional interaction between metabolic activities and the brain. So that it has been theorized the connection between mood, food, or 'the psychobiotic revolution': the brain health and state of mind are related to microbiome, a population of microbes living inside the intestines (Anderson et al., 2017).

It is not possible here to account for the multifaceted dynamics of a process such as metabolism, nor is this the aim of the article. However, we are already in the condition to understand that in metabolism can be seen an intelligent behavior, albeit implicit. In metabolism we detect typical traits of intelligence: creativity—in the sense of adaptative behavior—information, and memory. Traditionally referred to art, creativity is basically the capacity to give rise to new and unexpected phenomena, links, and relations. In different words, but with similar meaning, there are no rules that can prescribe creativity. In metabolism the main resource for a successful adaptation to environmental changes is creativity: organisms not to perish find new paths.

Extreme environmental conditions allow metabolism to find effective adaptive responses. One can only be surprised by the plasticity with which this occurs in nature. Take the *Drosophila melanogaster*, for instance. The fruit fly is a master of food-feeding flexibility when nutrients are scarce, because the «larval growth period is extended to allow additional growth and to ensure an appropriate final adult size under unfavorable growth conditions» (Koyama et al., 2020: 4524). The key role here is played by hormones (insulin, peptides with glucagon-like function, and steroid hormones) that systemically exchange information and provide the other biochemical components with specific signals. Such a behavior makes scientists aware of the 'flexibility' of the *Drosophila*'s metabolism that «*must be able to sense* and respond to changes in external environmental conditions and their internal state» (Koyama et al., 2020: 4523. *Italic is ours*).

The recent global alert for the climate change has thrown light on how fishes adjust their metabolic activity in warm water. How does it happen? Scientists investigated the behavior of *Perca fluviatilis* near a nuclear power plant in Sweden, where water (5–10°) is warmer than in other environments. Although warming is known to be highly dangerous for the fish's metabolism «by causing a mismatch between oxygen demand and supply, and a consequent reduction in aerobic scope (AS) and performance» (Ripley et al., 2023: 1), it has been evidenced a lower oxygen consumption, that is to say an unprecedent adjustment of the metabolic rate (Clark et al., 2013; Jutfelt, 2020).

This phenomenon leads, on a theoretical level, to a couple of considerations about implicit intelligence. First, we are in presence of an adaptive response that clearly transcends the biochemical dynamics of the organism itself; secondly, we are dealing with an emergence, an adaptive capability not given in ordinary conditions, which is the result of multiple interactions between metabolism and the environment, when a perturbation (thermal increase) occurs, and a new dynamic response (the effort to adapt to the environment) is effectively found.

As the previously sketched examples highlight, metabolic processes are informational at core, because each molecule is expected to communicate a specific content to organism. This duty is primarily performed by proteins, which are synthesized by metabolism. As Giuliani underlines:

> «Unlike other biological molecules such as sugars and lipids, which are identical in all living species, proteins are called 'informational'; their composition is encoded by sections of the DNA sequence and is peculiar to the individual species (...) The ordered course of metabolism strictly depends on the ability of the protein molecules to recognize the molecular actors of the specific chemical reactions to be promoted or inhibited» (Giuliani, 2022: 68. *The translation is ours*).

Finally, it should be noted the increasing importance of metabolic memory, as it appears with great evidence in the treatment of diabetes. For memory is intended the capacity, for the metabolic processes, to collect historical data about their activity, but also to keep track of the transformation occurred within the organic life, as a sort of living archive. All these information have been proved highly fruitful to decrease, for example, the risk if diabetic micro-and-macro vascular complication. Recently theorized, this topic is going to become a main topic for scientists, in terms of promising therapeutical perspectives. As Testa notes: «Recently, epigenetic mechanisms have been hypothesized to be a crucial interface between genetic and reduce the space between the two lines environmental factors to explain metabolic memory»[2] (Testa et al, 2017: 3). As it has already been noted for plants, living organism can be understood as knowledge accumulating systems «because they allow the most rapid and efficient responses to changes in environment» (Baluška & Mancuso, 2007: 205). If we translate this statement in our systemic view, living organism possesses and exhibits it in what they do, intrinsic intelligence.

3.3 *Cloud Computing, a Virtual Mind for Real Data*

As stated in the *Onlife Age* by Floridi (2015), in our age the real and the digital are melted together, and interconnections have become more and more strategically relevant, for the individual as well as for the organizations. The cloud computing

---

[2] *Epigenetics* is the study of how behaviors and environment can cause changes that affect the way genes work.

technology is properly and effective, as well as business-oriented way to interconnect the cyber-physical space.

A network digital infrastructure, cloud computing can be equated to an enormous system in which are stored data from public and private organizations, hospital, schools, universities, energy sources, banks, and many more. At least three primary services are hosted on the cloud, divided into infrastructure-as-a-service (IaaS), platform-as-a-service (PaaS), and software-as-a-service (SaaS). The remote data center—aka 'the cloud'—is interconnected with a network of users that, through the on-demand access can store their data, also having the possibility to exchange information with the other clients. As it has been noted: two important factors that trigger cloud computing systems (CCS) are reliability and energy efficiency (Dhawale & Dhawale, 2023). On the other side, a main concern is about the data sharing that «may contain some tactful or sensitive information». Thus, a main issue related to digital infrastructure deals with security and ethical concerns. However, another consideration to be done, which is also an area of interest in cloud computing, is the *autonomous capacity to operate.* We have reached the very heart of a leading infrastructure of the Fourth Revolution, in which the artificial intelligence is going to make a step forward.

> «It is expected that such resources can make independent decisions on their state to avoid human intervention in monitoring the state of these resources which have their direct and indirect cost implication monitoring regime toward the smart city applications» (Millham et al., 2023: 2251).

This possibility is far from remote. But before trying to understand why cloud computing is so promising in terms of *sui generis* intelligent thinking, it is worth noting how much the design of this digital infrastructure intertwines the human and virtual worlds. The operations center, the real cloud, recalls a brain in shape, while the network of the various users suggests the ramifications of the nervous system. The interactions between nodes (users) and the central system (cloud) evoke the plasticity of the messages transmitted from the center to the periphery and vice versa. Continuous inputs reach, like synapses, the stored data. As an example of a multiple system, cloud computing moves on at least three levels:

(1) man–machine interaction;
(2) the interactive processes between data;
(3) the emerging dynamics generated by the two previous interactions and by the dialogue, so to speak, of these new properties.

This is not science fiction. It already happens, as a consequence of the machine learning models, on which also the cloud computing technology is grounded. The infrastructure is expected to generate insights from big data. For instance, organizations can use machine learning to build models that automatically generate insights from big data, such as identifying trends or predicting future outcomes. Machine learning is a subset of artificial intelligence that involves the use of reduce the space between the two lines algorithms and statistical model to enable computers to 'learn' from data.[3] Nonetheless, one should not think of science fiction scenarios. There is a growing interest in developing artificial intelligence systems that are transparent and understandable or explainable AI, so that their decisions and predictions can be made available to humans.

What has been succinctly said about cloud computing opens a series of considerations in the light of this investigation.

It is important to underline that artificial entities are inherently intelligent, but their intrinsic intelligence, also in the case of artificial intelligence (AI) and machine learning (ML), results from the algorithms made by creative human beings. Compared to the processes that occur spontaneously in the world of life, they are the work of the intelligence that introduced them into the world as new phenomena. Thus, from this point of view, artificial entities are intelligent in an objective or intrinsic way, and in this sense the expression 'artificial intelligence' finds a proper and literal justification.

It is necessary to dispel a misunderstanding. Having conceived, and designed and engineered an artificial object, does not offer even to its creator any guarantee of full knowledge of the behavior of that object. Launched into the world, endowed with an autonomous dynamic, exposed to changes, the artificial object will often assume behaviors that are not only factually, but also theoretically unpredictable: artificial objects almost always have a considerable margin of self-organization and as the Golem of the Jewish tradition escape to varying degrees, even notably, from the intentions and expectations of those who designed and built them.

This observation can only advise us to adopt a principle of prudence: what we put into the world has unpredictable outcomes, and the unpredictable does not only include the favorable, but also the unfavorable, the destructive, and the dangerous.

4 Subjective—or Explicit—Intelligence

As noted before, explicit intelligence is the sort of human intelligence more investigated and definitely best known. Nor does it surprise, because it is directly observable through self-awareness, the exclusively human ability to take oneself as objects of observation and knowledge. Traditionally investigated by philosophy and logic, in recent times it has also become the object of other disciplines, including psychology and neuroscience. To explicit intelligence belong all the processes we directly know as part of our mental life, such as thoughts, reasoning, judgments, evaluations, choices, decisions, appreciations, as well as narratives, resolutions, projects, ideas, hopes, and much more. Subjective intelligence expresses itself in discoursive language, properly using both semantic and syntax rules. While sinking its roots in

[3] The use of the term 'learn' referred to digital machines is obviously metaphorical: Humans are said to learn because they are connected to their environment, both physical and cultural, while digital machines are said to 'learn' when thanks to their algorithm they select all the variables they have been exposed to during their training and find the best combination of these variables to solve a problem without being explicitly programmed.

the objective intelligence on which it depends and which feeds it, subjective intelligence is featured by its own traits, among which there is the constant search for the links that connect ideas and objects which are distant and disconnected at first sight.

The formal structure of inferences, regardless of the material content, has become the main topic of reasoning studies and explicit intelligence has quickly been identified with logical correctness. As a consequence, logic has assumed the powerful, but tacit and unproven premise, that the only correct inferences are those that can be formalized according to given rules and logicians set to work to formalize the rules of manipulation of symbols.

It follows that whatever cannot be formalized, i.e., most thought processes, is excluded from the investigation of human explicit intelligence. Most logicians have totally neglected that there are many fit and correct reasonings where conclusions do not necessarily follow from premises.

Peirce was the first in modern times to observe that the two main reasoning forms studied by logic, i.e., deduction and induction, are only small parts of our reasoning strategies. Furthermore, both show strong limitations: deduction produces only tautologies, and inductive reasoning limits itself to classifying similar cases under a general law. Peirce searching other arguments we reason by, identified a different inferential form, and called it 'abduction'.[4]

As Peirce writes, abduction is «the process of forming an explanatory hypothesis. It is the only logical operation that introduces some new idea». Abduction 'explains' a challenging issue by formulating the hypothesis that there is another fact, or a law, very different from the observed one—from which the observed one derives—which makes it possible an inverse inferential ascent, from effect to cause. The scientific method itself, in Peirce's words, begins with an abduction, formulating a hypothesis that explains an unexpected or surprising phenomenon and brings it back to normal by reorganizing the entire scenario according to a new perspective.

Logicians, who recently have begun to study abduction, have often tried to normalize it as the 'best explanation theory', which sounds as an attempt to get rid of the burden accounting for creativity, which exceeds logical rules and subverts the known formal order. Systemic thinking, being familiar with the emergences (II level systemic properties), which cannot be deduced from the first level ones, has appreciated abduction as a valuable cognitive tool for 'discovering' aspects, laws, entities not present in the available data.

Abduction is widely used by human intelligence in many domains: medical diagnosis, criminological investigations, politics, and also ordinary discourse, when one says to a friend 'He/she said he/she would do this, but I think that he/she has no intention to fulfill his/hers promise'.

Abduction, with deduction and induction, is one of the tools used by subjective intelligence to proceed in the comprehension of the world, but if we ask ourselves which is the main and proper characteristic of subjective intelligence, its common

[4] Aristotle in *Prior Analytics* uses the sintagm '*apagoghé*' to refer to the syllogism whose conclusions are 'believable', though not formally necessary.

seal, we should point at our meta-level capacity, as commonly experienced in self-awareness, the special moment where we take ourselves as objects under observation and also at work when we observe and describe the world. In philosophical jargon we speak of 'objectivation' to refer to such meta-level capacity.

Although this type of knowledge, concerning the self and the world, is limited and partial, it remains the highest product of explicit intelligence, which has been transmitted to us and which we can transmit in the forms of literature, poetry, the arts, the sciences, and philosophy.

5 The Emergence of Global Human Intelligence

As we summarily sketched in the previous paragraphs, objective and subjective intelligence could be thought in terms of the two faces of a same coin. In human beings their texture is so thick that they mutually recall each other. Quite differently, the sole objective intelligence does not allow to manage complex cognitive operations that must be supported with a language grounded on semantics and syntax. The animated beings that are endowed with only intrinsic intelligence lack of the ability to find reasons for their operations; it is impossible for them to share choices and to argue effectively in support of their activities and decisions.

Although it obeys to multiple constraints, objective intelligence does not have laws, nor rules or a moral system, neither shared culture, and not even literature, history, art, science, and philosophy. It would be impossible for human beings to even imagine a world devoid of the traits that most and best characterize us, in which we recognize ourselves.

A human being lacking intrinsic intelligence, on the other hand, does not even constitute him–herself as a human subject, because he/she is devoid of all the emergences that make him/her as such, starting from the bodily organization which requires a very high level of intelligence to constitute itself.

There is a significant asymmetry between our two main modes of intelligence: intrinsic intelligence can be maintained even in the absence or latency of the subjective one. Just think of sleep, states of brain-death, anesthesia. If in all these circumstances the body remains alive and vital, it is because its intrinsic intelligence is working, making animation possible. The reverse is not true: subjective intelligence simply fades when is not—or no more—preserved by the support of objective intelligence. Thus we can conclude that intrinsic intelligence has an ontological–anthropological primacy over subjective intelligence as this subjective property sinks its roots on the intrinsic intelligence.

It can be easy to realize how powerful consequences this conceptual frame is expected to have in bioethics, through the re-conceptualization of intelligent behavior in non-conscious states, like the coma. If a green plant—as briefly described before—has so many abilities to intelligently relate to, use and dominate its environment, how could its life be considered a not-life state? Thus, while terms like 'persistent vegetative state' (PVS) or post-coma unresponsiveness (PCU) should acquire a new meaning, this novel approach to 'intelligent behaviors' ought to bring about a change in the ongoing debate about the health care to patients in PVS or PCU.

A question arises at this point of the paper: what is the relationship between intrinsic intelligence and subjective intelligence? The problem is questioned by scientific literature in somewhat different terms, being the concept of intrinsic intelligence only recently proposed.

Concerning human intelligence, Ingold wonders whether the 'making' precedes the 'ideation' phase or it just follows it, if, in his words, 'making through thinking' precedes or follows 'thinking through making' (Ingold, 2013: 6). He concludes solomonically «making things is tantamount to a process of growth» (Ingold, 2013: xi).

If, in search for a neat answer, we turn to self-observation of artists, writers, philosophers, we nevertheless discover that they are often discordant in their answer to Ingold dilemma. Precisely, many authors report observing themselves in the act of making even in the absence of an idea or project explicitly present to their consciousness («I know what I'm saying only after I've said it…»).

Others, perhaps the majority, place conception before planning and making.

The fact that the evidence is discordant and heterogeneous does not allow reaching a so-called 'sociological' explication. As we suggest, a meta-level or conceptual clarification is needed.

In this case, introducing the concept of objective intelligence alongside the subjective one allows us to solve the dilemma: between the two conceptual frames there is no fixed antecedent relationship, in some cases the objective precedes the subjective one; in other cases, and without any scandal, it is true just the reverse.

Between the two terms referring to intelligence, there is a link of both interaction and interference, a kind of dance in which roles can be exchanged in the incessant process of our being in the world.

The continuous connection between intrinsic intelligence and subjective intelligence, according to which human intelligence emerges in its fullest sense, does not take place within a self-referential and solipsistic horizon.

The process of thinking originates in connection with what is other-than-us.

We call in question the environment.

We cannot treat the great topic of environment as it deserves, but we would at least warn from the mistake of thinking of the environment in a singular way: we are connected in multiple relationships and mutual transformation with multiple environments. In addition to the physical environment—in the multiple senses of chemical, biological, ecological, astronomical, architectural, economic environment, etc.—we are deeply influenced by the cultural, emotional, intellectual environment, and many others. We are aware of some of these connections, others transform us without us being able to grasp them explicitly.

In brief, we connect to them in a way that only rarely reaches consciousness and mostly happens at a subliminal level that, thanks to the activity of objective intelligence, let us participate in the fascinating and ever new process in which life consists of.

We project expectations of intelligent organization onto different environments, because only if the world in its many aspects is intelligent, we—who take a part in it—can hope to increase and ameliorate our knowledge of what surrounds us.

Finally, we can draw a conclusion from a systemic perspective: human intelligence is an emergent property of the human being system, which is structured thanks to the intertwining of objective intelligence, subjective intelligence and the different environments with which both are related. It is a skill of a high degree of complexity; consequently, there is no discipline that can claim to describe it exhaustively. Its various facets will be the subject of investigation by philosophy, logic, psychology, and neuroscience, but also by anthropology, sociology, economics, jurisprudence, medicine, and the arts, which will give their brushstroke to create the large fresco in which the 'human' narrates its self-representation.

A fresco to which every era adds its touch and to which we hope that this article will add a useful contribution to fade away problems created by obsolete paradigms and open and develop new perspectives in contemporary knowledge.

References

Anderson, S. C., Cryan, J. F., & Dinan, T. (2017). *The psychobiotic revolution. Mood, food, and the new science of the gut-brain connection.* National Geographic.

Anscombe, G. E. M. (1957). *Intention.* Blackwell.

Aristotle. (1924). In W. D. Ross (Ed.), *Metaphysics.* Clarendon Press.

Aristotle. (1949). In W. D. Ross (Ed.) *Prior and posterior analytics.* Clarendon Press.

Baldner, S. (2018). Thomas Aquinas and natural inclination in non-living nature. *Proceedings of the ACPA, 92*, 211–222.

Baluška, F., Volkmann, D., Hlavacka, A., Mancuso, S., & Barlow, P. W. (2006). Neurobiological view of plants and their body plan. In F. Baluška, S. Mancuso, & D. Volkmann (Eds.), *Communication in plants* (pp. 19–35). Springer.

Baluška, F., & Mancuso, S. (2007). Plant neurobiology as a paradigm shift not only in the plant. *Sciences Plant Signaling & Behavior, 2*(4), 205–207.

Baluska, F., Schlicht, M., Volkmann, D., & Mancuso, S. (2008). Vesicular secretion of auxin; evidences and implications. *Plant Signaling & Behavior, 3*, 254–256.

Biffi, I. (1974). Il giudizio "per quandam connaturalitatem" o "per modum inclinationis" secondo San Tommaso: analisi e prospettive. *Rivista Di Filosofia Neo-Scolastica, 66*(2/4), 356–393.

Błaszczyk, J. W. (2020). Energy metabolism decline in the aging brain-pathogenesis of neurodegenerative disorders. *Metabolites, 10*(11), 450. https://pubmed.ncbi.nlm.nih.gov/33171879/

Ceriello, A. (2012). The emerging challenge in diabetes: The "metabolic memory." *Vascular Pharmacology, 57*(5–6), 133–138.

Clark, T. D., Sandblom, E., & Jutfelt, F. (2013). Aerobic scope measurements of fishes in an era of climate change: Respirometry, relevance and recommendations. *The Journal of Experimental Biology, 216*, 2771–2782.

Clarke, A., & Fraser, K. P. (2004). Why does metabolism scale with temperature? *Functional Ecology, 18*(2), 243–251.

Darwin, C. (1880). *The power of movement in plants.* John Murray.

Darwin, C. (1891). *The movements and habits of climbing plants.* John Murray.

Derrible, S., Cheah, L., Arora, M., & Yeow, L. W., et al. (2021). Urban metabolism. In W. Shi (Ed.), *Urban informatics* (pp. 85–114). Springer.

Dhawale, C. A., & Dhawale, K. A. (2023). Review on reliability and energy-efficiency issues in cloud computing. In J. Wang (Ed.), *Encyclopedia of data science and machine learning* (pp. 790–802). IGI Global.

Drzewoski, J., Kasznicki, J., & Trojanowski, Z. (2009). The role of 'metabolic memory' in the natural history of diabetes mellitus. *Polskie Archiwum Medycyny Wewnetrznej, 119*(7–8), 493–500.

Fisogni, P. (2023). Machine learning and emotion. In J. Wang (Ed.), *Encyclopedia of data science and machine learning* (pp. 961–970). IGI Global.

Floridi, L. (2015). *The onlife manifesto. Being human in a hyperconnected era.* Springer.

Gagliano, M., Renton, M., Depczynski, M., & Mancuso, S. (2014). Experience teaches plants to learn faster and forgets lower in environments where it matters. *Oecologia, 175*, 63–72.

Gagliano, M., Vyazovskiy, V., Borbély, A. A., Grimonprez, M., & Depczynski, M. (2016). Learning by association in plants. *Scientific Reports, 6*, 1–9, https://www.nature.com/articles/srep38427

Giuliani, A. (2022). L'intelligenza del vivente. In L. Urbani Ulivi (Ed.), *L'intelligenza della natura* (pp. 47–89). EUPRESS FTL-Cantagalli.

Ingold, T. (2013). *Making: Archaeology, art and architecture.* Routledge.

Jutfelt, F. (2020). Metabolic adaptation to warm water in fish. *Functional Ecology, 34*(6), 1138–1141.

Keane, H. V. (1966). *Knowledge by Connaturality in St. Thomas Aquinas* [PhD dissertation, Marquett University]. Wisconsin.

Koyama, T., Texada, M. J., Halberg, K. A., & Rewitz, K. (2020). Metabolism and growth adaptation to environmental conditions in Drosophila. *Cellular and Molecular Life Sciences, 77*, 4523–4551. Retrieved online: https://link.springer.com/article/10.1007/s00018-020-03547-2

Legg, S., & Hutter, M. (2007). A collection of definitions of intelligence. *Frontiers in Artificial Intelligence and Applications, 157*, 17–24.

Libet, B. (2004). *Mind time. The temporal factor in consciousness.* Harvard University Press.

Manera, S. (2021). *Cervello intestino. Un legame indissolubile. Il ruolo del microbiota intestinale nelle nostre relazioni fisiche, mentali e psicoemozionali.* Macrolibrarsi.

Marconi, P. L., & Penna, P. (2021). The reaction of complex systems to symmetry braking events: A mathematical simulation model for neurosciences and social sciences. In G. Minati (Ed.), *Multiplicity and interdisciplinarity* (pp. 85–92). Springer.

McClintock, B. (1984). The significance of responses of the genome to challenge. *Science, 226*, 792–801.

Millham, R. C., Agbehadji, I. E., & Freeman, E. (2023). Nature-inspired algorithms and smart city applications. In Wang, J. (Ed.), *Encyclopedia of data science and machine learning* (pp. 2251–2270). IGI Global.

Minati, G. (2022). Sistemi: origini, ricerca e prospettive. Proprietà sistemiche emergenti e intelligenza. In L. Urbani Ulivi (Ed.), *L'intelligenza della natura* (pp. 91–127). EUPRESS FTL-Cantagalli.

Minati, G., & Pessa, E. (2006). *Collective beings.* Springer.

Peirce, C. S. (1857–1866). On the logic of science (Harvard lecture of 1865), Lecture II e Lecture VIII. In M. H. Fisch, C. Hartshorne, & P. Weiss (Eds.), *Writings of Charles S. Peirce* (Vol. I 1857–1866). Indiana University Press. *passim.*

Peirce, C. S. (1932). *Collected papers* (Vol. II, par. 636, 381). Harvard University Press.

Polanyi, M. (1966). *The Tacit dimension.* Doubleday.

Ripley, D. M., Quinn, F. A., Dickson, J., Arthur, J., & Shiels, H. A. (2023). Correction: Thermal preference does not align with optimal temperature for aerobic scope in zebrafish (Danio rerio). *Journal of Experimental Biology, 226*, 1–5.

Testa, R., Bonfigli, A. R., Prattichizzo, F., La Sala, L., De Nigris, V., & Ceriello, A. (2017). The "metabolic memory" theory and the early treatment of hyperglycemia in prevention of diabetic complications. *Nutrients, 9*(5), 437.

Torralba, J. M., & Llano, A. (2008). *Intention.* Conference paper held in Rome, at the Pontifical University of the Holy Cross, international conference about G.E.M. Anscombe, 28–29 February.

Torralba, J. M., & Llano, A. (2010). Representación y conocimiento pràctico. In *Thémata. Revista de Filosofía. Número* (Vol. 43, pp. 435–461).
Trewavas, A. J. (2009). What is plant behaviour? *Plant and Cell Environment, 32*, 606–616.
Trewavas, A. J. (2014). *Plant behaviour and intelligence*. Oxford University Press.
Trewavas, A. (2016). Intelligence, cognition, and language of green plants. *Frontiers in Psychology, 7*(588), 1–9. https://www.frontiersin.org/articles/10.3389/fpsyg.2016.00588/full
Urbani Ulivi, L. (2016). L'abduzione come momento della scoperta in contesti di realtà. Abduction is the inference that discovers a solution in problematic contexts. *Cassazione Penale, LVI*(5), 2240–2251.
Urbani Ulivi, L. (2022). Leggere il libro del mondo. In L. Urbani Ulivi (Ed.), *L'intelligenza della natura* (pp. 129–156). EUPRESS FTL-Cantagalli.
Whippo, C. S., & Hangartner, R. P. (2006). Phototropism; bending towards enlightenment. *The Plant Cell, 1*(8), 1110–1119.
Wuchty, S. (2014). Controllability in protein interaction networks. *Proceedings of the National Academy of Sciences, 111*, 7156–7160.
Zhang, Y., Liu, G., Yan, J., Zhang, Y., Li, B., &. Cai, D. (2015). Metabolic learning and memory formation by the brain influence systemic metabolic homeostasis. *Nature Communications, 6*, 6704. https://doi.org/10.1038/ncomms7704

The Complexity of Gaze

Pier Luca Bandinelli

Abstract The gaze is a conjugated movement of the eyes, which has evolved phylogenetically in the different species, with the initial and prevalent purpose of exploring the external environment. In the higher mammals, in addition to this function, in the course of evolution, the property of conveying and interpreting complex emotional states has been added, necessary for the communication processes between individuals, as well as for the construction of inferences on the mental states of others. The systemic interest of this topic consists in the fact that the mechanisms that regulate eye movements, and the multiple meanings of the gaze, are located in that subtle interface that separates neurology from psychiatry, materials that are at the same time indivisible due to the same substrate of origin, but infinitely distant on an epistemological level. In this theoretical reflection, both the complex neurological regulation and the different functions of eye movements, such as postural adjustments, will be analyzed, as well as the psychopathological meaning and the phenomenological description of the gaze in psychiatry, in an attempt at a holistic view of the phenomena described. The systemic conceptualization of this work, and its reference to multiple systems, consists in the fact that the theme of the gaze can be analyzed through two theoretical models that belong to different disciplines, but which interface in the construction of a common phenomenology, almost as if a point of view continually fluctuating between the two systems should be acquired.

Keywords Visual system · Gaze · Adaption

1 Phylogenesis and Embryogenesis of the Visual System

The visual system as a whole is one of the main interfaces between the organism and the environment and has developed over the course of phylogeny in its various signs (visual acuity, color discrimination, perception of movement, depth of field, speed of gaze, adaptation to light, etc.), initially to make the movement of the organism in

---

P. L. Bandinelli (✉)
Dipartimento Salute Mentale ASL Roma 1, Via Giovanni Martinotti 20, Rome, Italy
e-mail: pierluca.bandinelli@aslroma1.it

G. Minati and M. Pietronilla Penna (eds.), *Multiple Systems*, Contributions to Management Science, https://doi.org/10.1007/978-3-031-44685-6_7

its environment effective, basically for the search for food and the ability to escape from predators.

Only in the last phase of the evolution of mammals, to these characteristics was added that of being able to convey and interpret even very complex emotional states, up to the meta-representative capacity and the theory of mind (ToM).

The current enormous complexity of the synaptic network which, as a whole, forms the anatomical basis of the connections between the eyes and the brain, is the result of a very slow evolutionary process of millions of years, largely random, of trial and error in the interactions between nervous system and environment.

From a phylogenetic point of view, a first response to light was that of the phototaxis of some microorganisms, such as protozoa, which consisted in the possibility of being able to direct themselves toward a light source by these elementary life forms.

In a subsequent evolutionary line, the most rudimentary visual system belongs to a known group of marine arthropods, the trilobites, starting from the lower Cambrian up to the upper Permian (a period of time of 325 million years). The eyes of these animals are compound eyes, like those of insects, while the second line of development is that of vertebrates and is characterized by the chambered eye.

Both systems, albeit with a different level of complexity, provide a basic sensory structure that processes visual information at a first level of discriminative capacity (retinal cells), and subsequent neural stations, necessary for some reflex-type activities, and which then end in the occipital cortex, where the process of discrimination and integration of the received signal takes place and therefore of recomposition of the object seen. In turn, the occipital cortex, in addition to the great complexity of its intrinsic neural structure, has distributed connections with a high degree of functional parallelism with other cortical areas.

According to the well-known law of E. Haeckel whereby ontogeny recapitulates phylogeny, it is interesting to underline how already from the 18th day of embryonic life, the primary optic vesicle develops on the neuroectodermal wall of the future forebrain, connected to the diencephalon through the optic peduncle. With the growth and invagination of the vesicle, two layers are determined (internal and external) between which the retinal space develops. Around the 40th day the first draft of the lens begins to form, the complete growth of which will be completed around 20 years of age.

In the meantime, the pigmented layer of the retina originates from the external layer of the optic vesicle, while from the internal one, the nervous layer of the same, with the end of the process of differentiation between the different neural layers toward the 7th month.

From this moment, the eye is already sensitive to light, but the fovea, which represents the point of maximum visual acuity of the retinal layer, will differentiate only after the 4th month from birth.

From these descriptions, it is already clear, both as regards phylogenetic and ontogenetic processes, that the first function of the visual system is that of a simple discrimination and orientation of the organism toward light.

In a second time, and probably only as emergent properties of such a complex and dynamic system, both those much more integrated and complex visual activities that are typical of vertebrates, and additional unforeseen functions, will develop, and related to the exchange and discrimination of complex emotions signals.

2 The Neural Basis of Vision

The retina represents a highly complex and organized sensory epithelium, composed of specialized cells such as cones and rods, and a more properly neural, layered organization of bipolar cells, ganglion cells, and the network of interneurons.

The cones are responsible for daytime vision and accurately capture details and colors. The rods, on the other hand, convey a less sharp image, but are much more sensitive to light and allow the eye to see even in low light conditions.

The information that is processed by the rods and cones is then transferred to the bipolar cells which are connected to the ganglion cells.

Ganglion cells are divided into M (connected with a large number of rods and cones and sensitive to the movement of an object and rapid changes in the light context) and P (connected with a smaller number of receptors and specific in providing information on shape and color of an object).

Both M and P cells allow, through the subsequent integration capacity of interneurons, the formation of the so-called receptive field which is created by the complex interaction of continuous variations in the activation of ganglion cells with center-on (responding to light stimuli located in the center of their receptive field) and off-center ganglion cells (which respond to light stimuli located in the periphery of their receptive field).

The dynamic interaction between the two types of cells, mediated by the action of interneurons, favors a precise perception of contrasts and an adequate perception of changes in lighting. The retina, from a functional and structural point of view, has both a "blind area" located near the fibers that constitute the emergence of the optic nerve and the "fovea" the point of maximum visual acuity and both spatial and comparison discrimination chromatic. And it is for this reason that our gaze (coordinated by the oculomotor muscles) is in continuous movement, to allow the foveal area to analyze a particular observed, while at the same time building the meaning of the context that represents its frame.

The axons of the ganglion cells of the nasal and temporal part of the retina of each eye form the two optic nerves and which at the level of the optic chiasm partially cross their fibers to continue in the optic tract. Such a peculiar arrangement of the nerve fibers and the optic tract will result in different visual deficits depending on the level of a lesion. A lesion of the optic nerve will cause blindness in the visual hemifield ipsilateral to the lesion, while a lesion of the optic tract will result in a contralateral homonymous hemianopsia, i.e., blindness of the same half of the visual field of each eye. The optical tracts continue their journey toward the occipital cortex, but establishing connections with some brain nuclei for reflex or processing activities between visual information not yet fully integrated, with other cortical areas:

(1) The *pretectum nuclei* (in the mesencephalic-thalamic transition area) which regulate both pointing reflexes in response to a visual stimulus located peripherally to the foveal fixation point, and the intrinsic motility of the eye muscles, by varying the pupil diameter and the curvature of the lens;
(2) The *superior colliculus*, located in the midbrain, represents a nucleus of subcortical integration of visual information, which has not yet reached its full discrimination. It has a 7-layer histological organization and represents an input and output station of information that receives and projects to various cortico-subcortical areas (pulvinar, reticular formation, basal nuclei, and medial and anterior region of the parietal cortex) with the function of representation and somato-sensory, auditory, attentional, and proprioceptive integration. The superior colliculus represents a complex system of coordinate transduction, resulting in the integration of different sensory signals in the complex perception of the space explored by the senses. Furthermore, due to its projections to the nuclei of the base and from the parietal cortical areas, it coordinates the mechanisms of visual attention and gaze fixation, respectively.
(3) The *lateral geniculate body* (LGB), located in the thalamus, represents a processing station for the information conveyed by the optic tract, before reaching the occipital striatal cortex. A single lateral geniculate body receives afferents from the ipsilateral temporal hemiretina and the contralateral nasal hemiretina, i.e., two representations of the contralateral visual hemifield. Furthermore, the fibers coming from the individual retinas are composed of axons coming from both M-type ganglion cells and P-type ganglion cells. In this nucleus there is a level of integration of visual information which regards color contrast, luminance contrast, and the spatial and temporal frequency.
(4) From the *optical radiation* that starts from the LGB, it reaches the *primary visual cortex* located at the level of the calcarine fissure of the occipital lobe. It represents an area of very high integration and discrimination of the visual signal that works with high parallelism and through a specific columnar cyto-architecture. The information that comes from the LGB with respect to the contrasts between the images allows the extraction of the edges of the objects. *Simple cells* receive luminosity information by organizing them in a central (excitatory) and peripheral (inhibitory) area to form a new perceptual field which allows the recognition of a specifically inclined line, positioned in the excitatory area. The *complex cells* receive the information processed by the simple cells, and each perceptual field is placed side by side and partially superimposed so as to form areas without a specific activation/inhibition zone, but specific for edge recognition, according to a specific inclination regardless from the position in the perceptual field. *Blob cells*, on the other hand, are specialized in processing chromatic information, but totally insensitive to directions.

3 The Neurological Mechanisms of Gaze: Extrinsic Ocular Motility

In addition to the specific aspects of visual information processing, in this paragraph I will describe in a nutshell what are the neural basis of environmental exploration, but also of the expressive and emotional meanings, which depend on the different types of gaze. The cranial nerves responsible for the different gaze positions are the 3rd (oculomotor), the 4th (trochlear or pathetic), and the 6th (abducens).

The nucleus of the *oculomotor nerve* (3rd) is located in the midbrain, anterior to the aqueduct of Sylvius, and innervates all the extrinsic muscles of the eye (including the levator palpebrae) with the exception of the superior oblique and lateral rectus. It also contains visceral fibers for the intrinsic musculature of the ciliary muscle and the sphincter of the pupil. The *superior rectus* causes an elevation-only movement only when the eye is abducted by 23°; it has pure internal rotator function when the eye is abducted; in all intermediate positions, it has a mixed function of rotator and elevator. The *medial rectus* rotates the globe inward, bringing the cornea medially, in the horizontal plane (adduction). The *inferior rectus* rotates the eye downward when it is abducted; it is a pure external rotator when the eye is adducted and has a mixed function in the intermediate positions. The *lateral rectus* mainly rotates the globe outwards, bringing the cornea laterally, in the horizontal plane (abduction). The *inferior oblique* is a pure elevator in adduction; in abduction it is an external rotator so that the upper end of the vertical meridian leans out; furthermore, it draws the posterior half of the bulb downwards and therefore the cornea rises; finally, it draws in the posterior half of the bulb and then abducts it. The nuclear complex receives fibers from the ipsi- and contralateral cerebral cortex via the geniculate bundle of the internal capsule and from the superior colliculi via the tectobulbar bundle. It is also connected with the other nuclei of the oculomotor nerves, with the vestibular nuclei and with the nucleus of the spinal accessory nerve (XI) through the medial longitudinal fasciculus which extends up to the first 5 myelomeres of the cervical cord.

The nucleus of the *trochlear or pathetic nerve* (4th) is located in the midbrain at the level of the inferior colliculi, under the 3rd nuclear complex with which it shares the same input and output connections. It innervates the *superior oblique* and is a depressor when the eye is adducted 50°; in abduction it is an external rotator so that the upper end of the vertical meridian leans inward. It also lifts the posterior half of the bulb, and thus, the cornea rotates outwards. Overall rotates, lowers, and abducts the bulb.

The nucleus of the *abducens nerve* (6th) is located in the floor of the 4th ventricle. It shares the same incoming and outgoing connections as the other two cores. It innervates the *lateral rectus* and mainly rotates the globe outwards, bringing the cornea laterally, in the horizontal plane (abduction).

More generally, and closely connected both to environmental exploration and to the different types of emotional expression, different types of conjugated gaze movements can be distinguished which are the result of influences and control of different cortico-subcortical areas. Saccadic movements are the most important rapid eye movements, which have the function of bringing objects of interest that appear in the peripheral part of the retina to the fovea (horizontal controlled by ponto-mesencephalic nuclei; vertical and torsional controlled by the mesencephalic reticular formation). Other areas that influence saccadic movements are the superior colliculi, basal nuclei, cerebellum (with the vermis controlling the saccade metric and the flocculus controlling signal integration), the frontal lobe (with the prefrontal oculomotor area controlling controls predictive, memorized, and antisaccade saccades; supplementary oculomotor area which controls programming of saccades into memorized sequences; dorsolateral prefrontal cortex which inhibits reflex saccades and is involved in preparing memorized saccades), parietal lobe (with the area 7A which has the function of activating the attentional system and the lateral intraparietal area which contains neurons that are activated in the saccadic movement), the visual cortex of the intermediate and median area of the temporal lobe (which controls slow pursuit movements). The conjugated movements of the gaze, as well as from these areas described, are also controlled by a brainstem structure called the medial longitudinal fasciculus (FLM), which in turn receives afferents from the prefrontal oculomotor area and, within which cross fibers run which connect the 6th and 3rd, allowing the lateral rectus contraction of one side and the medial rectus of the contralateral eye.

4 The Gaze in Psychiatry

The mechanisms described above represent a synthetic basis that explains the neural bases of both visual discrimination processes and of the different gaze characteristics. At another level of analysis of the phenomenon, psychiatry focuses its interest on its meanings related to the different psychopathological declinations of the gaze which can be summarized as follows:

(1) In the first place, the ability to be able to discriminate and be able to describe with great accuracy, the different meanings of a look in relation to the basic emotional states (the infinite declinations of their transitions, their very specific differentiations, and possible confusions), characteristic this bankruptcy, e.g., in patients suffering from schizophrenic psychosis who tend to interpret some basic emotions incorrectly, in a more compromised way disgust, fear and surprise (Barkl et al., 2014), or interpret as threatening expressions that in normal people are perceived as neutral (Mitrovic et al., 2020). In these cases, some patients, precisely because of this discriminatory incapacity, can attribute an erroneous identity to people deemed similar to family members, but with the characteristics of impersonators, imposters and malevolent toward them (Coltheart & Davies, 2022).

(2) The interpretation of a look in its basic characteristics, in relation to the contextual frame in which it is determined, also represents an indication of the predictability of a behavior and can induce inferences on the mental states

or intentionality of others, also in relation to a certain social context. In this passage, it is highlighted how the gaze, together with other mimic characteristics of the face, or other postural and verbal signals, gives information both related to Theory of Mind (ToM) and skills in social cognition (Hiser et al., 2018). In particular, these authors describe how a circuit involving cortical areas (ventromedial prefrontal cortex) and subcortical areas (ventral striatum, amygdala, hippocampus, periaqueductal gray, and dorsal anterior cingulate cortex) is responsible for regulating decision-making processes, the generation and modulation of negative emotions, and the discrimination of facial emotional expression and related metacognitive abilities.

(3) The very complex meanings that evoke the different types of gaze in the psychiatrist who observes the patient during an interaction (the semiotics of the gaze). During an interview with a patient, observing the latter's gaze from a psychiatrist is one of the fundamental elements that determine some decisions, such as that of hospitalization in a psychiatric environment. The infinite declinations, and the hypothetical meanings of a look that can be described by a psychiatrist, can be innumerable. First of all, it is necessary to describe whether and how the patient maintains his gaze toward his interlocutor, or not. If it supports it, it is important to describe whether it is excessively persistent, if you constantly fluctuate between the gaze of the other and the environment, and in this case which areas of the environment are explored and with which emotional tone. It can be described as a worried or frightened expression that is combined with a look that explores around itself with circumspection as in paranoia, and it can be described as staring into space or at the ground, and, if accompanied by other expressive signs, related to sadness. It can be described as an elusive gaze that generally expresses vagueness or falsehood, or even a lively gaze, continuously in motion, combined with a very open eyelid rhyme and which expresses an elevation in mood.

5 The Systemic Point of View

The previous paragraphs have reported both the neural mechanisms that coordinate the various positions of the gaze, its meanings in determining a shared and biunique language between people, in terms of analysis of emotional meanings, subsequent metacognitive inferences, and of social cognition. And this both for his expressive abilities that can be grasped by the other toward us, often unaware of the person who expresses them, and for his discriminative abilities with respect to the analysis, elaboration and interpretation of his emotional meanings, and social, when it is interpreted by oneself toward the other.

It would be a reductionist thought to think that it is due to the neural circuits that the mechanism of the gaze is determined, which instead is generated by an intentional activity of the human being, in his "being-in-the-world", and in his relationship with the world.

An extremely interesting aspect of these extended characteristics of the gaze is that they are a property added during the evolution, beyond the primitive one and relating to the possibility of obtaining food and escaping predators. The evolution and differentiation of meanings from the primordial function, of survival in the environment, to that of representing an extremely complex relational system (both on a neural and psychological level), would seem to represent a characteristic relating to the concept of "systemic emergence". Systemic emergence represents an explanatory category of reality applicable to phenomena such as the mind and social changes, in opposition to ontological dualism and reductionism. The visual system, in the complexity with which it has evolved and due to its structure, has been co-constructed in a continuous interaction between the nervous system and the environment. So vision, mediated by gaze, and all its added properties that go beyond mere survival, is, in my opinion, the emergent property of the visual system. In the same way that our mind can be considered an emerging phenomenon of intentionality, in a systemic model that overcomes dichotomous visions between phenomenological and neuroscience-related paradigms (Bandinelli et al., 2013).

In this sense, in analogy with the visual system, it seems plausible that intentionality related to biological survival (food supply and escape from predators) has been among the most powerful facilitators and builders of our biological neural network since we were at the stage unicellular, and whose subsequent emergence was represented by the slow build-up of mental activity. All this process in the course of our evolution, like a continuous one-to-one oscillation of mutual interaction mind/brain—structure of the visual system/quality of gaze.

Furthermore, the diachronic and synchronic dimension of the emergence (Nagel, 1968), respectively, explain both the emergence over time of new physical/biological structures starting from pre-existing physical structures and the emergence of unpredictable phenomena.

The global unity of a complex system imposes organizational constraints on the parts (Morin, 1983), and the concept of "downward causation", whereby emergent properties have causal effects that also influence what happens at the underlying level, consists in the fact that the phenomenal aspects of the gaze can exert a causal action on the physical and biological systems to which they are associated, in a relationship of reciprocal influence. Therefore, every interrelationship endowed with a certain stability or regularity takes on an organizational character and produces a system which, considered as a whole, retroacts causally on its components to the advantage of the organization.

The systemic vision of the world must be able to represent an education and a continuous exercise for our mind to learn to oscillate between apparent reductionisms and dichotomies between different descriptive systems of reality (in the case of this work the same theme of the gaze as co-belonging both to psychiatry and neurology). An error of our thinking is to choose a descriptive cut of the real by eluding the other possible ones, as if one were to prevail over the other. Our mind, from the point of view of a systemic vision, must be educated to a continuous oscillation between descriptive paradigms, even antinomic ones, to allow us to have a vision of things that is both clear and out of focus.

References

Bandinelli, P. L., Fiocchi, A. M., & Ducci, G. (2013). La mente come fenomeno emergente dell'intenzionalità: La psicopatologia della schizofrenia in un modello sistemico oltre la dicotomia tra fenomenologia e neuroscienze. *Rivista Sperimentale di Freniatria, CXXXVII*(1), 131–148.

Barkl, S. J., Lah, S., Harris, A. W. F., & Williams, L. M. (2014). Facial emotion identification in early-onset and first-episode psychosis: A systematic review with meta-analysis. *Schizophrenia Research, 159*(1), 62–69. https://doi.org/10.1016/j.schres.2014.07.049

Brodal, A. (1983). *Neuroanatomia clinica*. Edizioni Ermes.

Coltheart, M., & Davies, M. (2022). What is Capgras delusion? *Cognitive Neuropsychiatry, 27*(1), 69–82. https://doi.org/10.1080/13546805.2021.2011185

Hiser, J., & Koenigs, M. (2018). The multifaceted role of the ventromedial prefrontal cortex in emotion, decision making, social cognition, and psychopathology. *Biological Psychiatry, 83*(8), 638–647. https://doi.org/10.1016/j.biopsych.2017.10.030

Mitrovic, M., Ristic, M., Dimitrijevic, B., & Hadzi Pesic, M. (2020). Facial emotion recognition and persecutory ideation in Paranoid Schizophrenia. *Psychological Reports, 123*(4), 1099–1116. https://doi.org/10.1177/0033294119849016

Morin, E. (1983). *Il metodo. Ordine, disordine , organizzazione*. Feltrinelli.

Nagel, E. (1968). *La struttura della scienza*. Feltrinelli.

Nieuwenhuys, R., Voogd, J., & van Huijzen, C. (1978). *The human central nervous system. A synopsis and atlas*. Springer.

Systemic Cognition: Sketching a Functional Nexus of Intersecting Ontologies

Rasmus Gahrn-Andersen, Maria S. Festila, and Davide Secchi

Abstract In pursuing a systemic view on cognition (as introduced by Cowley and Vallée-Tourangeau in Cognition beyond the brain. Springer, pp. 255–273, 2013), we recognize that cognition unfolds not just brain-side as postulated by traditional cognitive science (see e.g., Searle, The rediscovery of mind. MIT Press, 1992) but also world-side (cf. p. 255). This paper presents an attempt at specifying the systemic nature of human cognition by exploring it as occurring in the functional intersection of different ontologies (cf. Gahrn-Andersen in Organizational cognition: The theory of social organizing. Routledge, 2023). Part of the motivation for doing so is to inform views that overplay the uniqueness of experience (e.g., Jesus, Phenomenol Cogn Sci 17:861–887, 2018) and, thus, tone down the fact that cognition is imbued with and, hence, constrained by non-unique (or general) factors such as concepts, habits, instincts, affordances, activity trails etc. which ensure systemic coherence in the first place. More specifically, we pave new grounds for a systemic take on cognition by tracing human cognitive activity to what proponents of performativist Science and Technology Studies term 'practical ontologies' (Jensen, Berliner Blätter 84:93–104, 2021). Such ontologies have proven to be fruitful for disentangling messy human—non-human relations which are structurally co-determined by practices and their material arrangements. At the same time, however, little has been said about how such ontologies link up with aspects of the cognitive.

R. Gahrn-Andersen (✉) · M. S. Festila · D. Secchi
Department of Culture and Language, History and Communication, Faculty of Humanities, University of Southern Denmark, Søndre Stationsvej 28, 4200 Slagelse, Denmark
e-mail: rga@sdu.dk

M. S. Festila
e-mail: marfes@sdu.dk

D. Secchi
e-mail: secchi@sdu.dk

G. Minati and M. Pietronilla Penna (eds.), *Multiple Systems*, Contributions to Management Science, https://doi.org/10.1007/978-3-031-44685-6_8

1 Introduction: Cognition as Ontology-Performances

In the concluding chapter of their book, *Cognition beyond the brain*, Cowley and Vallée-Tourangeau (2013) introduce the notion of systemic cognition while explicitly recognizing that cognition unfolds not just brain-side (as postulated by traditional cognitive science, see e.g., Marr, 1982; Searle, 1992) but also in relation to the environment (p. 255). This means that 'drawing on human artifice, thinking is co-constituted by speech, movement and gesture. People distribute control as they link routines, make instant judgements and coordinate as they act' (p. 256). The description falls in line with classic distributed cognition work (e.g., Hutchins, 1995) while, at the same time, emphasizing synergies among resources which are irreducible to basic embodied interaction. By highlighting the deep interconnectivity that characterizes all resources involved in cognition, they suggest it shows features of a system.

The systemic nature of human cognition can be further specified in the sense that it is traceable to how our 'thinking' happens in the intersection of phenomena that are traditionally characterized as belonging to distinct ontologies (cf. Gahrn-Andersen, 2023, p. 39). Indeed, the emphasis on 'control'—e.g., who/what is the driver of any cognitive process—would be an ineffective way of approaching the study of cognition. Even in the case in which this perspective is pursued, it would only be the tip of the iceberg when exploring the systemic nature of cognition. This is simply because human cognition generally connects psychic or mental phenomena with other kinds of phenomena (e.g., biological, linguistic, social and material ones)—traditionally, all of these different phenomena have been traced to different, stand-alone ontologies which have been studied in separation. Maturana (2002) tacitly recognized this by tracing the basis of human cognition back to its biological roots in basic organismic activity and, hence, it has been called *bio-logic* (cf. Raimondi, 2019). As Maturana lengthily puts it:

> We human beings exist in structural coupling with other living and not living entities that compose the biosphere in the dimensions in which we are components of the biosphere, and we operate in language as our manner of being as we live in the present, in the flow of our interactions, in our domains of structural coupling (Maturana, 2002, p. 27).

The view resonates with other positions in Radical Embodied Cognitive Science (e.g., Chemero, 2009) that assume a strong continuity between living agents and cognition. Thus, as enactivists programmatically formulate it, 'where there is life, there is mind' (cf. Kirchhoff & Froese, 2017).

While pursuing a systemic take on cognition, however, one is necessarily bound to ask about how the general relates to the singular or to the unique or, more precisely, how a system's dynamics and structures interplay with individual experience (e.g., perceiving, imagining, reflecting). The move is important because it goes a step further from the basic insights provided by Cowley and Vallée-Tourangeau (2013) namely that cognition can be studied as a system where intrinsic and extrinsic factors intertwine in synergy. In this particular connection, it is their intersection which is central and specifically *how* such an intersection allows for cognition to unfold in a systemic fashion. The need for such an inquiry is exemplified by how De Jesus (2018) challenges anthropocentrism in cognitive science while emphasizing the need for going beyond naturalistic doctrines. In so doing, he acknowledges ontological multiplicity. In this connection, he argues in favor of us fully embracing the fact that organisms bring forth their "own 'unique' [ontological] worlds" (cf. p. 861) through their acts of sense-making. Such 'bringing forth' is ultimately traced to Francisco Varela's catchphrase: that enacted cognition can be metaphorically understood as being similar to 'laying down a path while walking' and, more specifically, organisms' 'distinct perceptual systems' (cf. p. 872) which enable them to be connected with their surroundings in the first place. The view critically targets the enactivist position that there might be different epistemic perspectives (or possibly experienced worlds) but that organisms nevertheless share the same reality.

2 Practical Generalities and the Seeming Uniqueness of Experience

Practical ontologies are particular constellations of concepts where research is still wide open to scholarly investigation. As such, a practical ontology

> makes explicit that the existence of one or many worlds will be the effect of interactions between innumerable heterogeneous actors that either create unity, multiplicity, or something yet different (Gad et al., 2015, p. 82)

As such, it is non-essentialist and builds on the recognition of the fact that cultural representations do not solely construct realities but must interplay with both the material and the practical (ibid., p. 75). The exploration of practical ontologies is based on the assumption that 'realities are negotiated, somewhere in-between, or to the side of, dualisms like ›fact‹ and ›fiction‹' (Jensen, 2021, p. 95). Moreover, there is a strong link to performativity in the sense that they are deemed to be relationally constituted meaning that practical ontologies do not exist as stand-alone and pre-fixed. Also, this explains the focus on plurality (i.e., ontologies). The basic claim is thus, as Law formulates it, that '*realities* (including objects and subjects) and *representations* of those realities are being enacted or performed simultaneously' (Law, 2008, p. 635). In connection with practical ontologies, it makes sense to place emphasis on systemic generalities as sources of, not necessarily stability, but, more importantly, *cohesion* across practical arrangements, modalities, etc. With generalities we mean the multiplicity of factors that are functionally determining of a cognitive system; such factors are necessary in order for the system to unfold as a coherent and recurrent practice. As such systemic generalities provide a degree of structure to how actions are performed, how many steps there are and how they relate to each other, and what to expect next (we will return to the theme of cohesion in Sect. 3).

Generalities are not solely extrinsic to human agents in the sense that they structure our actions strictly from the outside thus exerting a mere indirect influence on our experiences. Rather, as Hegel (1977[1807]) observed, although there is uniqueness to our perceptual experiences in the sense that they appear vastly rich and non-conceptual (a point also emphasized by Dretske, 1995), we find that there is nevertheless a degree of generality present—even in plain perception. So, although humans have traits of what Hegel terms 'natural consciousness' (i.e., conceptually unmediated experience) there is always more to our cognition in the sense that our experiences are permeated with, for example, social significance. Natural consciousness thus construed entails an immediate (hence, unmediated) receptive attitude toward elements of the environment which are then perceived in their indeterminate richness. Yet, natural consciousness as such remains a practical impossibility in both animal and human cognition given how general factors such as instincts (in animals) or concepts (in humans) shape our experiences. In the Hegelian sense, we enact general, yet concrete concepts (or what he terms *universals*) onto what appears in our phenomenological field. Even in the case of animals there is no natural consciousness as such in that they cannot simply stand passively over-against things, rather they exhibit a relation driven by a natural instinct (cf. Hegel, 1977, §109, p. 65) or 'affective compulsion' (Cussins, 2012, p. 28). In the case of human socio-practical dealings, one of us has emphasized in previous work (Gahrn-Andersen, 2021a) the role of concept-driven behavior and how the attribution of (tacitly 'labeling') a concept onto a 'thing' or 'object' plays a constitutive role to social practices. Specifically, it grants agents the means for identifying (through their performance) practically useful 'things' in accordance with the practical understandings of other agents partaking in the same praxis (Gahrn-Andersen, 2021a). A well-worn example in this connection is found in how Hutchins (2005) traces the possibility of collective queuing behavior to the fact that people have the common capacity of being able to conceptually identify a line of people as a 'queue'. What allows for such identification is the fact that the people involved evoke the understanding-driven comportment—in the form of what Heidegger (2010) termed an *as-structure*—onto the people standing in line who are thus perceived *as* a queue (cf. Gahrn-Andersen, 2021a). And, since what makes a queue is different in different cultural contexts (see Fig. 1 as the suggested 'queueing' in Milan Malpensa Airport), it is clear that even simple conceptualizations require practice-oriented understandings.

On a systemic view, however, it is insufficient to locate generalities that impinge on or structure individual experience solely agent-side. Moreover, there is also a pitfall related to intellectualizing practical behavior in the sense of assuming that human agents themselves are perfectly able to provide informative accounts on their practical performances by means of the explications of concepts and reasons. As such, discursive elements are not solely defining of praxis (for an example, see Højbjerg, 2012 in Gad et al., 2015, p. 72). As such, there is a need for acknowledging core insights from performativist Science and Technology Studies (STS) and Actor Network Theory (ANT). Here, the basic assumption that agency is irreducible to human or other living agents is particularly relevant. Surely, this does not mean that we should grant inert objects or infrastructures 'life.' Instead, we can assume there is agency in the sense of acknowledging that even non-living entities have the potential to influence us through the practical arrangements (assemblages) they form part of (cf. Latour, 1993). On an ANT-view, human—non-human relations are

Fig. 1 Queue signs at Milan Malpensa Airport

symmetrical in the sense that there is co-influence at stake thus rendering it sensible to ascribe agency to non-human—and even non-living—phenomena. For instance, Jensen (2017) shows how leaky sewages, insufficient waste management, climate change, and other factors all influence the daily life of the inhabitants of Phnom Penh. The problem with run-down sewage system links with other material, social and environmental issues as well. In this connection, Jensen argues that 'pipe dreams' should not be seen as mere political or social 'wish rumors' but rather '*emanations* of infrastructural arrangements' (Jensen, 2017, p. 633). Jensen writes:

> Even if people at [Phnom Penh's] Kandal market rarely think about pipes, and even if their dreams, interpretations and critiques are about something else (like culture and politics), the capacity for having those dreams is consequent upon infrastructural activity trails (ibid., p. 645).

So, although the infrastructure itself gets visible every time the Japanese contractor in charge of replacing Phnom Penh's sewage pipes digs up a new stretch of road, it is generally the cascading effects (and causes) of a poorly functioning sewage infrastructure which are the concerns of Phnom Penh's inhabitants. Thus, it is indeed 'possible that Cambodian pipes can have real effects irrespective of whether the people they affect think so' (ibid., p. 645). In other words, the infrastructure itself becomes a nexus for the practical generalities (e.g., blocked roads due to maintenance works, sewage flows, waste disposal, corrupt politicians, frequent local floodings, etc.) to shape individual experience.

3 Systemic Cohesions, Generalities, and Trail-Making

It is always possible for an observer to infer various points of connection (and, hence, elements of cohesion) between the events, material structures, agents of different kinds, etc., that constitute a particular cognitive system. We emphasize the importance of uncovering a logic of cohesion as a means for connecting otherwise more or less speculative points (or fictions) concerning the nature of a given cognitive system and, hence, the relations between its different elements. Ultimately, we aim to explore how such a system unfolds as a practical *assemblage* reflecting a plurality of ontologies. Specifically, we seek to uncover systemic stability and, thus, avoid that the abolishing of the subject-object (or mind-matter) dichotomy ends up in a radical constructivist project as the one expressed by Jensen (2021) when clarifying Pickering's (2005) notion of 'dances of agency': 'Bypassing the distinction between mind and matter, we would be able to observe diverse agencies coming together in »dances of agency« that reciprocally tune and unpredictably modify all of them' (Jensen, 2021, p. 95). We use the terms 'system' and 'assemblage' interchangeably to underscore the ontologically heterogenous reality of human socio-cognitive relations.[1] Our approach takes its basis in the functional locus of the cognitive system under consideration and, more specifically, in the generalities that constitutionally contribute to the system's persistence and evolution.

We critically follow up on Jensen's (2017) take on pipe infrastructure and how it shapes so-called *activity trails* which impose order or disorder on, for instance, urban spaces (p. 632). As a term introduced on the basis of the abolishing of subject-object dichotomies, Jensen contrasts activity trails with phenomenological approaches 'that take human embodiment and sense-making capacities as their analytical ground', arguing that 'the notion of activity trails thus enables decentred analyses in which subject and object formations are emergent outcomes of material–relational processes' (p. 631). Yet, if we consider how Cussins (2012), the inventor of the notion, thematizes 'activity trails', we find that his account is more nuanced. One reason for this is that it is dispositions of the agent which are constitutive of the trails that emerge: 'Trails are in the environment, but their characterization depends on an animal's sensitive–active relation to the world' (Cussins, 2012, p. 27). More specifically, it is an agent's capacity for being sensitive to such trails that grants the trails their particular practical-material significance. Thus, there is without a doubt a degree of generality which conditions the *for*'ness of a given trail in the sense that 'a trail for an ant is not a trail for a monkey' (ibid.). But Cussins refers to mediational content as a mediator or 'solicitor' for certain activities in the subject. Rather, in remaining faithful to Cowley and Vallée-Tourangeau's non-representational and systemic focus, we instead place emphasis on the agent's competent understanding-driven comportment through *as*-structured (and, hence, the enactment of generalities

[1] One may argue that 'assemblage' is a particular kind of system where a number of components are either put together (by someone) or happen to be together (at random) and arrange themselves to form mutual relationships. For the purpose of the discourse developed in this chapter, we deem these arguments partially irrelevant.

such as concepts, skills, and habits/instincts) for the constitution of the activity trails which a given cognitive system—or practice—consists of.

3.1 Exemplar: Leakage Detection and Repair Case

In what follows, we present insights from a study on leakage detection and repair in the context of a large Danish utility company (see also, Gahrn-Andersen, 2020, 2021b; Cowley & Gahrn-Andersen, 2023). Ethnographic data was collected at the Maintenance Department through semi-structured interviews and observations during multiple visits between 2022 and 2023. We are particularly interested in uncovering how different activity trails influence the leakage detection practice as a cognitive system. In accordance with our focus on the coherence logics, we focus explicitly on the generalities which are vital for the upholding and persistence of the practice. Accordingly, we now turn to exemplifying how connections between different material configurations were constructed and performed when a leakage was identified in the basement of an old apartment building in a rather affluent district of Copenhagen in Denmark. This specific case not only involves many heterogeneous elements converging and diverging through the practice of leakage detection and repair, but also serves to show how the enactment of systemic generalities such as engineering terms, mundane concepts ('green water', 'pipes', and 'basement'), software representations, refined technical skills, and the acquired habits of experts give cohesion to the way activity trails intersect and gain practical-material significance (Fig. 2).

The utility company owns part of the building's pipe network and hence the right and obligation to access and maintain it. Residents have private possessions in storage that are kept under lock, but which obstruct the district heating workers' access to the pipes. Although filled with many private possessions, the space itself is not

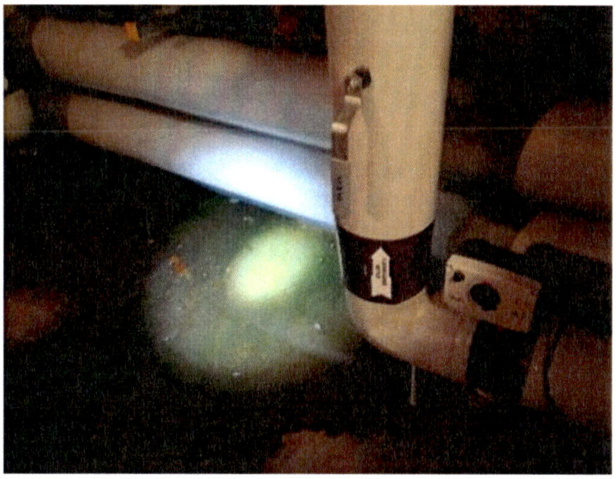

Fig. 2 Green water of district heating (chemically altered by the utility company to be distinguishable from drinking water when leaking from pipes) in the basement of an apartment building in Denmark

privately owned, yet it is not a public space. It is the rather peculiar material configuration of pipes, valves, meters, locks, moving boxes, knick-knacks, and old furniture that grants rights and enforces obligations, constraining practice-specific cognitive phenomena such as concepts, skills, and habits to be brought to the fore by the district heating workers in their quest to stop the leaking green water. Thus, the basement is a hub for a bundle of different activity trails and the ontologies they sustain which nevertheless all somehow influence the unfolding of the observed leakage detection/ repairing practice. Indeed, the basement is simultaneously performed *as* a private storage space by the residents, *as* an administrative responsibility by the building's management, *as* a network access point by the infrastructure engineer, *as* a coordination problem for the case responsible (e.g., 'Cases [in this area] are always exciting because… I don't want to sound too rough, but some of the people think they are a little privileged'), *as* a site of jurisdictional negotiation between the utility company and the building's administration,[2] and *as* the space for enacting repair work for diggers, welders, and technicians. The basement is thus a site of many overlapping practices and, thus, quite different yet general understanding-driven behavior. Indeed, the basement's practical significance far exceeds the standard denotative meaning of 'a basement' as 'the part of a building that is wholly or partly below ground level'.[3] As we unfold the story of this leakage repair case, we return to these different socio-material configurations to show how the district heating workers' competent understanding of heterogeneous worlds grant them particular practical-material significance in the context of leakage detection and repair.

3.2 *Different Ontologies at Work*

On a cool Danish summer day, one rather uneventful for district heating workers (as is the majority of summer days when hot water consumption is less demanding on the network), one of us was visiting the maintenance department office of the utility company. In the midst of sporadic computer work, two of the maintenance engineers received a confirmation that they now have access to the basement of an apartment building, where a suspected hot water leakage was reported the previous day. On their previous attempt to investigate the leakage, the engineers realized they did not have access to the whole piping area, as part of the basement was used as private storage by the building's residents and hence locked up. As one engineer explained:

> We are going on private property, and that is why I was stopped yesterday with the work… because I do have the permission to cut a lock and get access to my piping, even though it's private property… not inside people's houses, but in the storage space I can cut the lock and just enter if I wanted to, but it means a lot of paperwork and stuff […]. Now we have talked to the janitor to break the lock, and they put some plastic zip straps that we can cut with the pliers.

---

[2] In the sense that the heat exchange system in the basement marks the boundary between the district heating infrastructure and the building's heating system.

[3] https://www.merriam-webster.com/dictionary/basement-accessed April 9, 2023.

Even before the engineers reached the site of the leakage, the practical significance of the socio-material configuration of the basement was brought forth by the engineers. In this connection, the need to access the basement was overplayed by the engineers' understanding of private property, procedural rules and ultimately, by their habit of avoiding unnecessary paperwork. Here, systemic generalities stemming from heterogeneous socio-material practices (e.g., the laying of the piping, paperwork requirements, basement accessibility, various kinds of ownership, ambiguous jurisdictional status of the leakage site) shaped the way in which the engineers experienced the leakage site and imposed a certain logic to how the leakage repair unfolded: It was indeed wiser to ask the janitor to break the lock—he would not need to do paperwork.

As the janitor confirmed he had broken the lock and placed plastic zip straps instead, the engineers were 'back on the case'. Before heading to the address specified on the leakage referral, one of them accessed the Geographic Information System (GIS) that depicts the utility company's heating pipe network and the general technical specifications of that area of the network. He did so in order to get a detailed overview of the kind of pipes and components in the basement. Specifically, he typed in the address provided in the referral and zoomed in to the component codes matching the location and then created a PDF with the technical drawing of the affected area for printing (see Fig. 3).

In doing so, the engineers engaged with generalities as he got an idea of the valves nearby (that must be closed, should pipe exchange work be involved), or if there was any history of breakdowns in the specific area, and how they had been handled in the past. In this connection, the engineers used the GIS to extend their cognitive activities by representing aspects of the pipe network which is otherwise be inaccessible to their basic embodied perception. This enabled them to make sense of the general configuration of the pipes and components which might be affected by the leak. As Fig. 3 shows, this involved making sense of the material configuration of the basement from a standpoint involving technical concepts in preparation of on-site

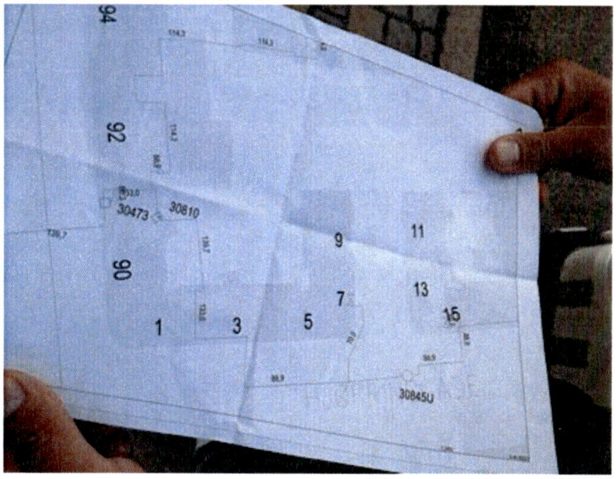

Fig. 3 Engineer holding a printout of a technical drawing from the GIS

investigation. This concept-driven cognition was enabled by the engineers' professional experience and skillful usage of artefacts and technologies (e.g., GIS). Indeed, systemic generalities such as ordinary and technical concepts, GIS-representations, habits, and skills enabled the engineers to relate to the site of the leakage and enact its interwoven material arrangements even before gaining physical access to the site.

On-site, individual experience such as the perception of environmental cues (e.g., sight of green water, rusty pipe, high humidity, and high temperature) interplayed with concept-driven cognition (e.g., knowing that all these phenomena are cues for district heat water leaking from the pipes) and enabled the engineers to identify the reported leakage as being indeed the utility company's responsibility. This identification enabled the following activity trail to unfold through the leakage detection practice: using a thermographic camera to identify the exact location of the leakage. To have access to the entire system of pipes, the engineers needed to access private storage rooms and move boxes and furniture around. Moreover, they habitually used their expert knowledge that pipes usually break at the bents and also followed the water trails to identify that the leakage was coming from a very rusty pipe inside the wall. Following the successful identification of the leakage, the engineers could easily assign the digging and welding contractors needed for carrying out the repair. Once detecting the potential source of leakage, they put the private belongings back where they were found, locked the basement with a new plastic strip, and drove to the office where they made a note in the GIS for the contractors for what they needed to do.

4 Conclusion

The utility company engineers unfold the leakage detection practice as trail-making activity which brings order to the chaos of a basement filled with green water and to a dysfunctional district heating network. In this process, the engineers retort to various generalities in the form of technical skills and knowhow, material artefacts (e.g., particular types of pipes and valves), technical and mundane concepts, past experience with leakage detection, prejudices about people living in an affluent neighborhood, the GIS, etc. in order to successfully navigate the different activity trails that culminate in the basement. Our account shows that systemic cognition is functionally sustained by individual perception, team-coordination, and the performance of aspects pertaining to different intersecting practical ontologies which normally have very little do with one another (e.g., personal storage, a janitor's plastic strips, hot water, a printout of a PDF based on GIS information, and paperwork aversion) but which in the context of the particular leakage come to gain functional relevance as the engineers use their expertise to conduct leakage detection and repair.

Acknowledgements All authors acknowledge the funding awarded by the Velux Foundation of the project Determinants of Resilience in Organizational Networks (DRONe) (grant number 38917).

References

Chemero, A. (2009). *Radical embodied cognitive science*. MIT Press.

Cowley, S. J., & Gahrn-Andersen, R. (2023). How systemic cognition enables epistemic engineering. *Frontiers in Artificial Intelligence, 5*, 960384.

Cowley, S. J., & Vallée-Tourangeau, F. (2013). Systemic cognition: Human artifice in life and language. In S. J. Cowley, & F. Vallée-Tourangeau (Eds.), *Cognition beyond the brain* (pp. 255–273). Springer. https://doi.org/10.1007/978-1-4471-5125-8_14

Cussins, A. (2012). Environmental representation of the body. *Review of Philosophy Psychology, 3*, 15–32.

De Jesus, P. (2018). Thinking through enactive agency: Sense-making, bio-semiosis and the ontologies of organismic worlds. *Phenomenology and the Cognitive Sciences, 17*, 861–887.

Dretske, F. (1995). *Naturalizing the mind*. MIT Press.

Gad, C., Jensen, C. B., & Winthereik, B. R. (2015). Practical ontology: Worlds in STS and anthropology. *Nature and Culture, 3*, 67–86.

Gahrn-Andersen, R. (2020). Making the hidden visible: Handy unhandiness and the sensorium of leakage-detecting drones. *The Senses and Society, 15*(3), 272–285.

Gahrn-Andersen, R. (2021a). Conceptual attaching in perception and practice-based behavior. *Lingua, 249*, 102960.

Gahrn-Andersen, R. (2021b). Transcending the situation: On the context-dependence of practice-based cognition. In T. Ciecierski & P. Grabarczyk (Eds.), *Context dependence in language, action, and cognition* (pp. 209–228). De Gruyter.

Gahrn-Andersen, R. (2023). Perception as conceptual orderliness in enlanguaged practices. In D. Secchi, R. Gahrn-Andersen, & S. J. Cowley (Eds.), *Organizational cognition: The theory of social organizing*. Routledge.

Hegel, G. W. F. (1977). *Phenomenology of spirit*. Oxford University Press.

Heidegger, M. (2010). *Being and time*. SUNY Press.

Højbjerg, C. K. (2012). Den 'ontologiske vendings' utilstrækkelighed. *Tidsskriftet Antropologi, 67*, 29–35.

Hutchins, E. (1995). *Cognition in the wild*. MIT Press.

Hutchins, E. (2005). Material anchors for conceptual blends. *Journal of Pragmatics, 37*, 1555–1577.

Jensen, C. B. (2017). Pipe dreams: Sewage infrastructure and activity trails in Phnom Penh. *Ethnos, 82*(4), 627–647.

Jensen, C. B. (2021). Practical ontologies redux. *Berliner Blätter, 84*, 93–104.

Kirchhoff, M. D., & Froese, T. (2017). Where there is life there is mind: In support of a strong life-mind continuity thesis. *Entropy, 19*, 169. https://doi.org/10.3390/e19040169

Latour, B. (1993). *We have never been modern*. Harvester-Wheatsheaf.

Law, J. (2008). On Sociology and STS. *The Sociological Review, 56*(4), 623–649.

Marr, D. (1982). *Vision: A computational investigation into the human representation and processing of visual information*. W.H. Freeman.

Maturana, H. R. (2002). Autopoiesis, structural coupling and cognition: A history of these and other notions in the biology of cognition. *Cybernetics & Human Knowing, 9*(3–4), 5–34.

Pickering, A. (2005). Asian Eels and global warming: A posthumanist perspective on society and the environment. *Ethics and the Environment, 10*(2), 29–43.

Raimondi, V. (2019). The bio-logic of languaging and its epistemological background. *Language Sciences, 71*, 19–26.

Searle, J. (1992). *The rediscovery of mind*. MIT Press.

Complex Processes

Mechanisms Underlines Brain Processes in Addiction: A Spiking Neural Network Analysis from the EEG

Roberta Renati, Natale Salvatore Bonfiglio, and Maria Pietronilla Penna

Abstract Research in the field of pathological addiction has shown that craving is one of the factors contributing to continued substance use. Craving can be considered an urgent need to use substances and can be triggered by a prolonged period of abstinence or by external stimuli. Several studies have shown how the use of non-invasive neurostimulation techniques such as tDCS (Transcranial Direct Current Stimulation) can help reduce craving through the use of protocols and repetitive stimulation sessions. The present study aims to analyze the neurological mechanisms underlying the brain processes, stimulated by neurostimulation, through the use of Spiking Neural Network (SNN), which represents a biologically more adherent method compared to traditional neural network models, for learning, classification, and comparative analysis of brain data. EEG traces were recorded on eight subjects with cocaine dependence during a cognitively activating task. Four subjects performed 9 sessions with tDCS and the remaining 4 subjects did not perform tDCS. Anode placed on F3 and cathode on F4 has been used as tDCS protocol. EEG data were collected through a Brain–Computer Interface (BCI). The NeuCube architecture, trained on spatio-temporal EEG data, was used to classify EEG data between the two groups of subjects to facilitate a more in-depth comparative analysis of dynamic processes in the brain. The results showed different brain activity patterns, in particular related to the frontal areas. Through the analysis of the connectivity of the SNN model and its dynamics, new information has been revealed on the effects of tDCS on cognitive functions of the brain and areas involved in cocaine addiction. These results can bring new insights into the brain functions associated with brain activity in subjects with addiction under cognitive activation conditions.

R. Renati · M. P. Penna
Department of Pedagogy, Psychology, Philosophy, University of Cagliari, Cagliari, Italy
e-mail: roberta.renati@unica.it

M. P. Penna
e-mail: penna@unica.it

N. S. Bonfiglio (✉)
Statistical Service, IRCCS, Centro San Giovanni di Dio Fatebenefratelli Brescia, Brescia, Italy
e-mail: nbonfiglio@fatebenefratelli.eu

G. Minati and M. Pietronilla Penna (eds.), *Multiple Systems*, Contributions to Management Science, https://doi.org/10.1007/978-3-031-44685-6_9

1 Introduction

Addiction is a chronic condition characterized by impulsivity, loss of control, and compulsion in seeking and using a substance, leading to the emergence of negative emotional states when access to the substance is denied (Koob & LeMoal, 2005). Some authors support the idea that there are common neurochemical mechanisms involved in substance use linked to disorders of the reward system: the central reward pathway involves structures of the dopaminergic system such as the mesolimbic cortical of the ventral tegmental area and projections to the nucleus acubens and prefrontal cortex.

One of the main symptoms in individuals with substance use disorders (SUD) is craving, defined as a pressing, urgent, and irrepressible desire to use substance, which in most cases results in a loss of control (O'Brian, 2011), remaining a central symptom of the addiction process. Regular substance use and addiction status are commonly related to the influence of stimuli (cue) related to the substance. The sight or smell of an alcoholic beverage or the context of dealing or consuming the substance, for example, increase physiological arousal and craving (Bonfiglio et al., 2019), which contribute to the maintenance of addictive behavior and increased likelihood of relapse. Given the wide range of environmental stimuli that trigger acute craving, it seems likely that craving is elicited by conditioning processes, following a specific causal pathway (Sadok et al., 2008). Therefore, episodic craving has been often reproduced in the laboratory with the use of paradigms such as exposure to stimuli that recall substances (cue exposure) to study its underlying mechanisms.

Recently, the limitations of pharmacological use on reducing craving are leading to experimentation with alternative options, such as non-invasive neurostimulation techniques (Nitsche et al., 2003). Transcranial direct current stimulation (tDCS) is an effective non-invasive brain stimulation technique in which a weak current is applied to the brain through the use of electrodes, allowing modulation of brain activity based on the polarity of the current. Specifically, anodal stimulation increases cortical excitability, while cathodal stimulation decreases it (Nitsche et al., 2003, 2000; Priori et al., 1998; Lupi et al., 2017). Non-invasive brain stimulation, such as tDCS, has been successfully applied for the reduction of impulsivity and craving in patients with substance dependence. The most common protocols used aim to reduce craving for a substance by applying direct current specifically to the prefrontal area, for the regulation of electrical activity related to the reward system.

Many studies have analyzed the electrical activity using electroencephalography (EEG) both while performing a task (such as an attentional task) and in resting state on individuals with substance use disorders (SUDs) (Liu et al., 2022). These studies have also analyzed the alterations represented by neuropathology-related changes and represented a new way of examining functional brain alterations related to the chronic substance use (Shulte et al., 2014). Conventionally, EEG signals have been divided into the following frequency bands in the frequency domain, which have been associated with different cognitive processes: delta (0–3 Hz), theta (4–7 Hz), alpha (8–12 Hz), beta (13–28 Hz), and gamma (> 29 Hz). Oscillations in each band serve different functions depending on the location of their sources and parameters such as amplitude, frequency, and phase. For example, slower frequencies are thought to reflect the coordinated activity of large-scale neuronal networks, while higher frequencies may indicate local activity. The functional connectivity may refer to the synchronization of EEG rhythms in different brain areas, which includes direct and indirect connectivity, that can be represented as a network composed of nodes (i.e., regions) and edges (i.e., connectivity between regions) (He & Evans, 2010).

Moreover, EEG is able to capture spatio-temporal brain data (STBD) in an activated condition (i.e., while performing a cognitive task) with high temporal resolution and is able to detect changes in cortical activity that occur within milliseconds. Therefore, EEG is one of the most reliable techniques for measuring dynamic neurocortical changes associated with specific brain areas.

The EEG has been widely used for addiction studies, as it allows the effects of drugs on the brain to be measured, which often manifest themselves in the form of changes in post-synaptic potentials that correspond to alterations in EEG activity (Doborjeh et al., 2015). Indeed, the reinforcing effects of many drugs are precisely mediated by the mesolimbic dopamine pathway, which alters EEG tracings (Knyazev, 2007).

Recently, spiking neural network (SNN) models have been proposed as a reliable technique to record and process STBD, due to the closeness to the biological model than classical neural network models. Neurons communicate in the form of binary events called spikes and each neuron transfers and receives chemicals and produces electrical current. SNN models are, therefore, a good example inspired by highly parallel neuromorphic implementations of the brain (Kasabov et al., 2014; Paugam-Moisy & Bohte, 2012).

The aim is to analyze the effectiveness of tDCS in reducing craving and to analyze the brain areas most involved in the craving response. To do this, we will use EEG tracings that will be processed by an SNN model in order to identify areas and connections between areas in a biologically adherent manner. The hypothesis is that tDCS has an effect in reducing craving, resulting in a regularization of brain activity associated with the left and right dorsolateral area.

2 Methodology

Participants

Ten subjects admitted to a residential addiction treatment service were involved in the study. The subjects were selected on a voluntary basis, having been informed by the facility manager of their willingness to participate in the trial. The subjects included (a) had received a diagnosis for SUD for cocaine according to DSM5, (b) had been abstinent from substances for at least 30 days and were of legal age. Subjects with serious health problems (e.g., hepatitis), aged over 65 years or ineligible for electromedical and neurostimulation equipment (e.g., presence of metal plates, etc.) were excluded.

Out of 10 subjects involved, 8 were male. The subjects were divided into two conditions (with and without tDCS) and before and after the trial (T0 and T1). The mean age of the subjects was 38.2 (11.2). The subjects were all polyabusers with cocaine use as their main substance, had been abstinent for 41.4 (19.2) days, had started using at an average age of 22.3 (8.9) years. Four subjects had substitution therapy, 5 had psychopharmacological therapy for psychiatric symptoms, and 1 subject had no drug therapy.

Procedure

The apparatus consisted of a portable PC on which the images were presented in a random sequence. The EEG was registered by a 14 channel EEG BCI helmet (emotive+, www.emotiv.com). The helmet was assembled by an expert operator who subsequently started the software. The subject sat in front of the monitor at a distance of approximately 50 cm, with the Pc leaning on the table.

The experiments were performed in lab conditions and with no distractions. The 14 EEG electrodes were placed on the scalp according to the International 10–20 System for acquisition of raw data from electrodes that were positioned at AF3, F7, F3, FC5, T7, P7, O1, O2, P8, T8, FC6, F4, F8, and AF4 positions. Odd numbers of electrodes were positioned to the left hemisphere of the brain; even numbers of electrodes were positioned to the right hemisphere of the brain. Two referencing electrodes CMS (on the left side) and DRL (on the right side) were used for reduction of noise in signal. Signals were sampled with 128 Hz sampling rate and were sent through the wireless connection to the dongle, which is used as a receiver connected to the USB port of personal computer. A host computer was used to perform computationally expensive analysis of acquired data and transformation into the control signals for software applications (Vokorokos et al., 2012).

The trial consisted of the exposure, through the use of a specially built software connected to a 14 channels BCI helmet, of 30 images linked to substance and 10 neutral images, all presented randomly. Each image remained to the screen for 5 s. The images were chosen by each subject, together with the experimenter, to identify those that had a greater salience for the subject.

A tDCS protocol was applied before the trial session relative to the presentation of craving images, with the aim of reducing craving for the substance and increasing behavioral inhibition. We used the BrainStim machine (EMS S.r.l., Bologna, Italy), with two electrodes (5×5 cm) and sponges soaked in saline solution. We used the EEG 10–20 system to place the electrodes on the scalp.

For craving reduction, we applied a direct current of 2.5 mA for 20 min. The anode was applied on the left dorsolateral prefrontal cortex (left DLPFC; F3) and the cathode was placed on the right DLPFC (F4) (Coles et al., 2018). Subjects received tDCS session twice a week for five weeks, always before to begin trial with craving images.

The methodology for analyzing SNNs includes several algorithms for processing the EEG signal, based on the following modules.

(a) *Data Encoding*: Spatio-temporal EEG data are processed as temporal sequences of continuous real values, and subsequently converted (encoded) into discrete spikes. The threshold-based method for EEG encoding was used here. If the signal variation increases above a certain spike threshold at a consecutive time, a positive spike is generated. If, on the other hand, the signal decreases below a threshold, then a negative spike is generated. If neither of the previous two cases occur, no spike is generated.
(b) *Mapping*: EEG data are mapped into a 3D SNN scheme, using the Talairach brain model (Giacometti et al., 2014). Variables in the input EEG data were positioned in the SNN model as input neurons with respect to their spatial coordinates (x, y, z), based on those in the Talairach brain atlas.
(c) *Learning*: After the spatial mapping of the SNN model, unsupervised learning was performed using a spike time correlated learning rule (Spike Time Dependent Plasticity; STDP) (Masquelier et al., 2009). In this study, different SNN models were trained with EEG data for different conditions, e.g., before (time T0) and after (time T1) a tDCS protocol, in different groups of participants. The SNN models of T0 and T1 were subtracted to capture the differences between the two states.
(d) *Visualization of the Models*: To better understand the change in spatio-temporal interactions between brain areas in relation to the changes obtained with the tDCS protocol, the SNN models were visualized in a 3D space. We visualized 2 SNN models, each of which was trained separately with EEG data from each T0 and T1 condition.
(e) *Pattern Classification*: In order to perform a classification task, a classifier was trained, in supervised mode, to associate each subject with the relevant condition (pre and post, i.e., T0 and T1), and thus learn the association between the patterns and between the trained SNN connectivity and the class label information. In this case, the Dynamic Evolving Spiking Neural Network (deSNN) was used as a classifier (Kasabov et al., 2013).

3 Results

The data were processed with NeuCube software (Kasabov et al., 2013). The classification analysis was performed in a supervised mode, categorizing the subjects into two groups: (1) subjects who did not perform tDCS, that is, whose brain activity was not yet affected by stimulation with tDCS and (2) subjects who performed tDCS, that is, whose brain activity was still affected by stimulation with tDCS. All subjects at T0 were considered as belonging to group 1, while at T1 subjects who did not perform tDCS were considered as 1 and instead those who performed tDCS were considered as 2.

The network included a total of 1471 connections in the input layer, and 2 nodes (corresponding to the two subject categories) in the output layer. The analysis performed with NeuCube resulted in 80% correct classifications for both category 1

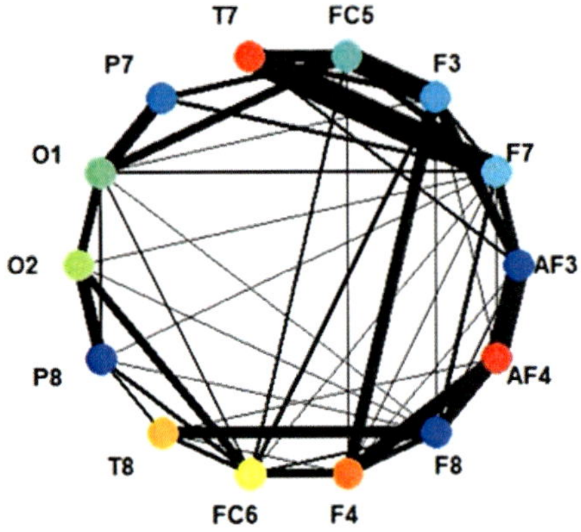

Fig. 1 Brain areas connections after learning for group 1

and category 2. The parameters used were as follows: Spike threshold (0.5), training set ratio (0.5), Training time length (1), Validation time length (1), Potential leak rate (0.002), STDP rate (0.01), Threshold of firing (0.5), Training round (2), refractory time (6), LDC probability (0), classifier = deSNNs (Mod = 0.8; Drift = 0.005; K = 3; Sigma = 1).

Group 1: Brain Activity Not Yet Stimulated with tDCS

Figure 1 shows the connections between brain areas in subjects whose brain activity has not yet been affected by stimulation with tDCS. The thicker the lines, the stronger the connection (average of connection weights) between areas. A strong connection can be seen between areas F3 and F4, F8 and T8, T7 and F7, as well as between the occipital and parietal areas and between all frontal and antero-frontal areas with their respective adjacent areas.

Figure 2 shows the transversal section of the network with the 14 points corresponding to brain areas. In red are the inhibitory connections and in blue are the excitatory connections. Thicker connections indicate a stronger connection between nodes. It can be seen that the connections are mostly diffuse, and no particular patterns can be distinguished.

Group 2: Brain Activity Yet Stimulated with tDCS

Figure 3 shows the connections between brain areas in subjects whose brain activity was still affected by stimulation with tDCS. The thicker the lines, the stronger the connection (average of the connection weights) between the areas. The absence of a strong connection between areas F3 and F4 can be seen, and also the reduction in the strength of connections between the occipital and parietal areas and between all frontal and antero-frontal areas with their respective adjacent areas.

Figure 4 shows the transversal section of the network with the 14 points corresponding to brain areas. In red are the inhibitory connections and in blue are the excitatory connections. Thicker connections indicate a stronger connection between

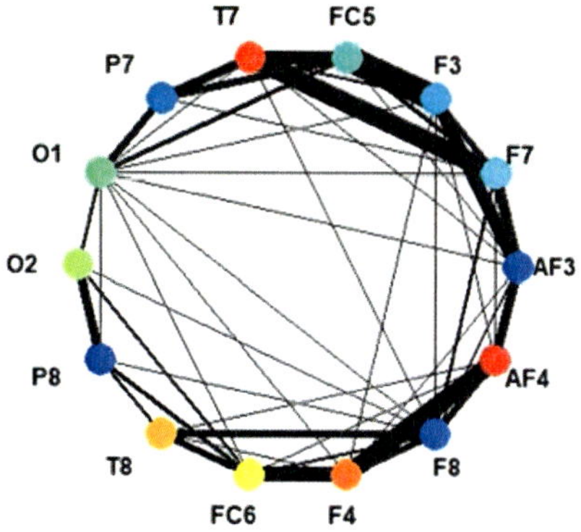

Fig. 2 Brain activity after learning (right) with reference position (left) for group 1

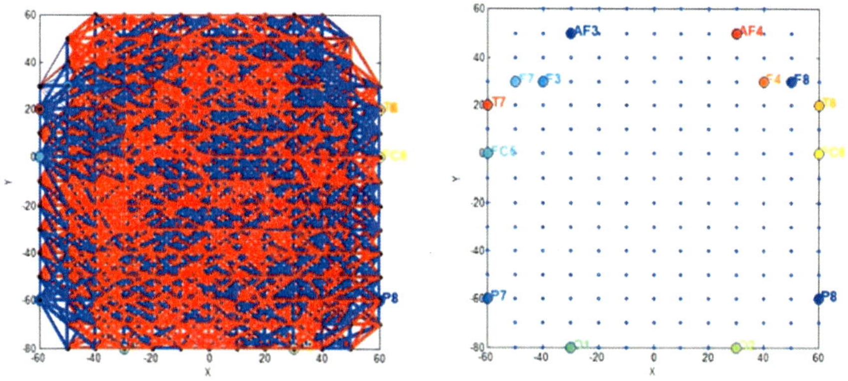

Fig. 3 Brain areas connections after learning for group 2

nodes. A much clearer pattern of connections can be seen than in Fig. 2, with a predominance of positive connections and high weights around the left dorsal lateral prefrontal cortex (DLPC) and right and left occipital areas. The same can also be seen for the right DLPC but a greater presence of inhibitory connections.

4 Discussion

The neurological mechanisms underlying the brain processes, stimulated by neurostimulation, using SNN, have been analyzed in this study, in a group of subjects with cocaine dependence.

What has emerged is the presence of a very different pattern of connections between subjects who had performed tDCS sessions and subjects who had not performed tDCS sessions. In particular, the brain pattern of the subjects who had

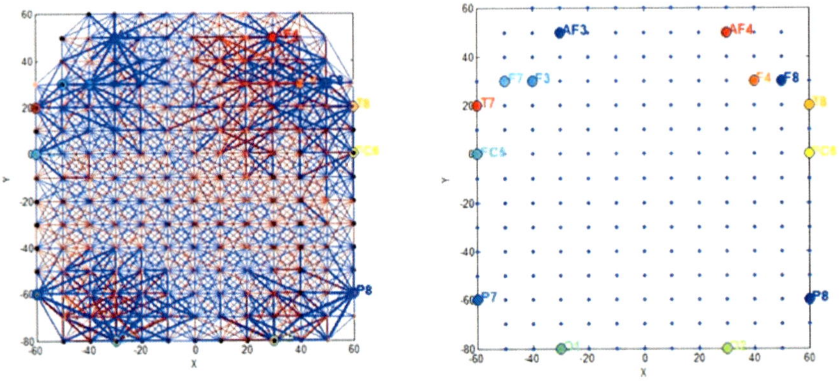

Fig. 4 Brain activity after learning (right) with reference position (left) for group 2

performed tDCS seems very confused and chaotic and unstructured, unlike the subjects who had performed tDCS. For those who had undertaken tDCS, we can see the involvement of some sites referring to particular networks such as the intersection between F3 with AF3 and F4 with AF4 and FC5 and FC6 (salience network: SN), F3 and F4 (default mode network: DMN, reward network: RN and central executive network: CEN), intersection between P7 with O1 and P8 with O2 (SN and DMN), O1 and O2 (DMN), and the intersection between F7 with T7 (RN).

There is a regularization of these networks in the comparison between group 1 and group 2 (see Figs. 1 and 3) and a decrease in the average weights of con-connections (Fig. 1) and a greater presence of inhibitory connections relatively to the anterior part of the DMN system (e.g., intersection F3/G4 and AF3/AF4) and the RN (F3/F4). With respect to the RN, the presence of greater inhibitory activity in the F4 area and the reduction in the average weight of connections between F3 and F4 confirms the finding in the literature that craving is associated with inhibitory activity of the F3 area (Bonfiglio et al., 2021; Mazzoleni et al., 2019).

The DMN comprises the posterior cingulate, posterior parietal cortex, and the ventromedial prefrontal cortex. The DMN is composed by a set of remote, functionally connected cortical nodes less active during executive tasks and more active in resting state. Therefore, its activity is *anti-correlated* with goal-oriented tasks and abnormal activity has been associated with several psychological disorders (Simon & Engström, 2015). It is implicated in Theory of Mind, episodic memory, and self-reflective processes.

The CEN includes the dorsolateral prefrontal cortex and posterior parietal cortex and is involved in cognitively demanding task requiring attention.

The SN comprises the ventrolateral prefrontal cortex and anterior insula (mutually referred to as the fronto-insular cortex) and the anterior cingulate cortex. The SN responds to the degree of subjective salience, whether cognitive, homeostatic, or emotional (Goulden et al., 2014).

Interactions between SN and the DMN are thought to be important for cognitive control (Jilka et al., 2014). The SN responds to external behaviorally salient events, whereas the high activity of DMN has been reported when subjects have an internal focus of attention, such as during internally directed thought (Buckner et al., 2008). Therefore, changing from automatic behavior—e.g., where attention is focused internally—to behavior guided by external events is accompanied by increased activation within the SN and deactivation of the DMN (Sharp et al., 2011). The right anterior insula, a key node of the SN, is thought to causally influence activity in DMN (Chiong et al., 2013).

The DMN has been seen to be correlated in multiple brain diseases, including substance use disorders (SUD) (Wang et al., 2016). Altered DMN function has been associated with rumination, emotional dysregulation, and compromised cognitive functions (Simon & Engström, 2015). Dynamic interactions between the DMN and other networks can influence cognition and emotion and affect attentional performance and impulsivity (Shannon et al., 2011). In SUD, both anomalous DMN function and disturbed interaction with other networks impair cognitive and affective processes and can contribute to craving and relapse.

Sutherland et al. (2012) have proposed a framework to describe the relation between the above-mentioned networks in the addicted brain. According to authors, SN would direct attention resources toward internal withdrawal symptoms (shifting brain functions toward DMN; and away from CEN), whereas under drug administration, the SN would direct attentional resources toward external stimuli and executive functions (shifting brain activity toward CEN and away from DMN). This switch between DMN and ECN is also suggested to be mediated by insula (Sutherland et al., 2012).

When exposed to drug-related stimuli, therefore, addicted individuals showed increased activity in anterior and posterior DMN (Zhang & Volkow, 2019). The resting functional connectivity of the anterior DMN, which contributes to the attribution of personal value and emotional regulation, tends to be decreased; whereas resting functional connectivity of the posterior DMN, which focuses attention to the internal world, tends to be increased. Abnormal resting state within the DMN is believed to contribute to impaired self-awareness, negative emotions, and to ruminations in addiction (Zhang & Volkow, 2019). Moreover, drug and cue-evoked increases in anterior DMN activity are positively related to craving and relapse in cocaine abusers, alcoholics, and nicotine dependence (Zhang & Volkow, 2019).

Ultimately, in this study, we have shown that tDCS over DLPFC in early abstinent cocaine users could enhance functional brain connectivity of RN and CEN through increased correlation between left DLPFC (the major hub of RN and CEN) and other functionally-related regions in this network and through a reduction connectivity with right DLPFC. Indeed, previous work has shown that regions of the brain which respond to the degree of subjective salience of a stimulus, the SN, are responsible for switching between the DMN and the CEN (Sridharan et al., 2008).

5 Conclusion

On the basis of our results, we can speculate that impaired executive function, including the capacity to exert self-regulation over emotions and desire for drugs (craving), facilitates rumination and obsessive thinking about drugs (during preoccupation phase) that leads to relapse and compulsive drug taking despite adverse consequences (Volkow et al., 2016). These findings may provide useful guidance for future research in the direction of analyzing brain activity through the use of neural networks and SNNs. They also provide input for professions that operationally structure interventions and treatment plans with patients for craving reduction and relapse prevention.

The limitation of the work lies in having used few subjects and the use of a helmet that analyzes EEG on only 14 channels. An analysis on a larger sample of subjects and the use of an EEG recording system on many more channels could surely provide more and much more detailed information about the functioning of brain networks and their relationship to the RN and CEN systems.

References

Bechara, A. (2005). Decision making, impulse control and loss of willpower to resist drugs: A neurocognitive perspective. *Nature Neuroscience, 8*(11), 1458–1463. https://doi.org/10.1038/nn1584

Bonfiglio, N. S., Renati, R., Agus, M., & Penna, M. P. (2019). Validation of a substance craving questionnaire (SCQ) in Italian population. *Addictive Behaviors Reports, 9*, 100172.

Bonfiglio, N. S., Renati, R., Patrone, L., Rollo, D., & Penna, M. P. (2021, June). The use of cognitive training with tDCS for the reduction of impulsiveness and improvement of executive functions: A case study. In *2021 IEEE international symposium on medical measurements and applications (MeMeA)* (pp. 1–6). IEEE.

Bouton, M. E. (2002). Context, ambiguity, and unlearning: Sources of relapse after behavioral extinction. *Biological Psychiatry, 52*(10), 976–986.

Buckner, R. L., Andrews-Hanna, J. R., & Schacter, D. L. (2008). The brain's default network: Anatomy, function, and relevance to disease. *Annals of the New York Academy of Sciences, 1124*(1), 1–38.

Cohen, S., Kamarck, T., & Mermelstein, R. (1983). A global measure of perceived stress. *Journal of Health and Social Behavior*, 385–396.

Coles, A. S., Kozak, K., & George, T. P. (2018). A review of brain stimulation methods to treat substance use disorders. *The American Journal on Addictions, 27*(2), 71–91.

Conklin, C. A., & Tiffany, S. T. (2002). Applying extinction research and theory to cue-exposure addiction treatments. *Addiction, 97*(2), 155–167.

Chiong, W., Wilson, S. M., D'Esposito, M., Kayser, A. S., Grossman, S. N., Poorzand, P., Seeley, W. W., Miller, B. L., & Rankin, K. P. (2013). The salience network causally influences default mode network activity during moral reasoning. *Brain, 136*(6), 1929–1941.

Doborjeh, M. G., Wang, G. Y., Kasabov, N. K., Kydd, R., & Russell, B. (2015). A spiking neural network methodology and system for learning and comparative analysis of EEG data from healthy versus addiction treated versus addiction not treated subjects. *IEEE Transactions on Biomedical Engineering, 63*(9), 1830–1841.

Dunsmoor, J. E., Niv, Y., Daw, N., & Phelps, E. A. (2015). Rethinking extinction. *Neuron, 88*(1), 47–63. https://doi.org/10.1016/j.neuron.2015.09.028

Giacometti, P., Perdue, K. L., & Diamond, S. G. (2014). Algorithm to find high density EEG scalp coordinates and analysis of their correspondence to structural and functional regions of the brain. *Journal of Neuroscience Methods, 229*, 84–96.

Goldstein, R. Z., & Volkow, N. D. (2002). Drug addiction and its underlying neurobiological basis: Neuroimaging evidence for the involvement of the frontal cortex. *The American Journal of Psychiatry, 159*(10), 1642–1652. https://doi.org/10.1176/appi.ajp.159.10.1642

Goudriaan, A. E., De Ruiter, M. B., Van Den Brink, W., Oosterlaan, J., & Veltman, D. J. (2010). Brain activation patterns associated with cue reactivity and craving in abstinent problem gamblers, heavy smokers and healthy controls: An fMRI study. *Addiction Biology, 15*(4), 491–503.

Goulden, N., Khusnulina, A., Davis, N. J., Bracewell, R. M., Bokde, A. L., McNulty, J. P., & Mullins, P. G. (2014). The salience network is responsible for switching between the default mode network and the central executive network: Replication from DCM. *NeuroImage, 99*, 180–190.

He, Y., & Evans, A. (2010). Graph theoretical modeling of brain connectivity. *Current Opinion in Neurology, 23*(4), 341–350.

Jansen, J. M., Daams, J. G., Koeter, M. W., Veltman, D. J., Van Den Brink, W., & Goudriaan, A. E. (2013). Effects of non-invasive neurostimulation on craving: A meta-analysis. *Neuroscience & Biobehavioral Reviews, 37*(10), 2472–2480.

Jilka, S. R., Scott, G., Ham, T., Pickering, A., Bonnelle, V., Braga, R. M., Leech, R., & Sharp, D. J. (2014). Damage to the salience network and interactions with the default mode network. *Journal of neuroscience, 34*(33), 10798–10807.

Kasabov, N., Dhoble, K., Nuntalid, N., & Indiveri, G. (2013). Dynamic evolving spiking neural networks for on-line spatio-and spectro-temporal pattern recognition. *Neural Networks, 41*, 188–201.

Kasabov, N., Feigin, V., Hou, Z. G., Chen, Y., Liang, L., Krishnamurthi, R., Othman, M., & Parmar, P. (2014). Evolving spiking neural networks for personalised modelling, classification and prediction of spatio-temporal patterns with a case study on stroke. *Neurocomputing, 134*, 269–279.

Knyazev, G. G. (2007). Motivation, emotion, and their inhibitory control mirrored in brain oscillations. *Neuroscience & Biobehavioral Reviews, 31*(3), 377–395.

Koob, G. F., & Le Moal, M. (2005). Plasticity of reward neurocircuitry and the 'dark side' of drug addiction. *Nature Neuroscience, 8*(11), 1442–1444.

Lee, J. L. C., Nader, K., & Schiller, D. (2017). An update on memory reconsolidation updating. *Trends in Cognitive Sciences, 21*(7), 531–545. https://doi.org/10.1016/j.tics.2017.04.006

Liu, Y., Chen, Y., Fraga-González, G., Szpak, V., Laverman, J., Wiers, R. W., & Richard Ridderinkhof, K. (2022). Resting-state EEG, substance use and abstinence after chronic use: A systematic review. *Clinical EEG and Neuroscience, 53*(4), 344–366.

Lupi, M., Martinotti, G., Santacroce, R., Cinosi, E., Carlucci, M., Marini, S., Acciavatti, T., & di Giannantonio, M. (2017). Transcranial direct current stimulation in substance use disorders: a systematic review of scientific literature. *The Journal of ECT, 33*(3), 203–209.

Masquelier, T., Guyonneau, R., & Thorpe, S. J. (2009). Competitive STDP-based spike pattern learning. *Neural Computation, 21*(5), 1259–1276.

Mazzoleni, M., Previdi, F., & Bonfiglio, N. S. (2019). Classification algorithms analysis for brain–computer interface in drug craving therapy. *Biomedical Signal Processing and Control, 52*, 463–472.

Minervini, I., Palandri, S., Bianchi, S., Bastiani, L., & Paffi, D. (2011). Desire and coping self-efficacy as craving measures in addiction: The self-efficacy and desire scale (SAD). *The Open Behavioral Science Journal, 5*(1).

Monfils, M.-H., Cowansage, K. K., Klann, E., & LeDoux, J. E. (2009). Extinction-reconsolidation boundaries: Key to persistent attenuation of fear memories. *Science, 324*(5929), 951–955. https://doi.org/10.1126/science.1167975

Nader, K., Schafe, G. E., & Le Doux, J. E. (2000). Fear memories require protein synthesis in the amygdala for reconsolidation after retrieval. *Nature, 406*(6797), 722–726. https://doi.org/10.1038/35021052

Nitsche, M. A., & Paulus, W. (2000). Excitability changes induced in the human motor cortex by weak transcranial direct current stimulation. *The Journal of Physiology, 527*(Pt 3), 633.

Nitsche, M. A., Liebetanz, D., Antal, A., Lang, N., Tergau, F., & Paulus, W. (2003). Modulation of cortical excitability by weak direct current stimulation–technical, safety and functional aspects. *Supplements to Clinical Neurophysiology, 56*, 255–276.

O'Brien, C. (2011). Addiction and dependence in DSM-V. *Addiction, 106*(5), 866–867.

Paugam-Moisy, H., & Bohte, S. M. (2012). Computing with spiking neuron networks. *Handbook of Natural Computing, 1*, 1–47.

Priori, A., Berardelli, A., Rona, S., Accornero, N., & Manfredi, M. (1998). Polarization of the human motor cortex through the scalp. *NeuroReport, 9*(10), 2257–2260.

Przybyslawski, J., & Sara, S. J. (1997). Reconsolidation of memory after its reactivation. *Behavioural Brain Research, 84*(1–2), 241–246. https://doi.org/10.1016/s0166-4328(96)00153-2

Shannon, B. J., Raichle, M. E., Snyder, A. Z., Fair, D. A., Mills, K. L., Zhang, D., Bache, K., Calhoun, V. D., Nigg, J. T., Nagel, B. J., Stevens, A. A., & Kiehl, K. A. (2011). Premotor functional connectivity predicts impulsivity in juvenile offenders. *Proceedings of the National Academy of Sciences, 108*(27), 11241–11245.

Sharp, D. J., Beckmann, C. F., Greenwood, R., Kinnunen, K. M., Bonnelle, V., De Boissezon, X., Powell, J. H., Counsell, S. J., Patel, M. C., & Leech, R. (2011). Default mode network functional and structural connectivity after traumatic brain injury. *Brain, 134*(8), 2233–2247.

Sadock, B. J., & Sadock, V. A. (2008). *Kaplan & Sadock's concise textbook of clinical psychiatry*. Lippincott Williams & Wilkins.

Schulte, M. H., Cousijn, J., den Uyl, T. E., Goudriaan, A. E., van den Brink, W., Veltman, D. J., Schilt, T., & Wiers, R. W. (2014). Recovery of neurocognitive functions following sustained abstinence after substance dependence and implications for treatment. *Clinical Psychology Review, 34*(7), 531–550.

Simon, R., & Engström, M. (2015). The default mode network as a biomarker for monitoring the therapeutic effects of meditation. *Frontiers in Psychology, 6*, 776.

Sridharan, D., Levitin, D. J., & Menon, V. (2008). A critical role for the right fronto-insular cortex in switching between central-executive and default-mode networks. *Proceedings of the National Academy of Sciences, 105*(34), 12569–12574.

Sutherland, M. T., McHugh, M. J., Pariyadath, V., & Stein, E. A. (2012). Resting state functional connectivity in addiction: Lessons learned and a road ahead. *NeuroImage, 62*(4), 2281–2295.

Tiffany, S. T., & Drobes, D. J. (1990). Imagery and smoking urges: The manipulation of affective content. *Addictive Behaviors, 15*(6), 531–539.

Vokorokos, L., Mados, B., Adám, N., & Baláz, A. (2012). Data acquisition in non-invasive brain-computer interface using emotiv epoc neuroheadset. *Acta Electrotechnica Et Informatica, 12*(1), 5.

Volkow, N. D., Koob, G. F., & McLellan, A. T. (2016). Neurobiologic advances from the brain disease model of addiction. *New England Journal of Medicine, 374*(4), 363–371.

Wang, H., Guo, W., Liu, F., Wang, G., Lyu, H., Wu, R., Chen, J., Wang, S., Li, L., & Zhao, J. (2016). Patients with first-episode, drug-naive schizophrenia and subjects at ultra-high risk of psychosis shared increased cerebellar-default mode network connectivity at rest. *Scientific Reports, 6*(1), 1–8.

Zhang, R., & Volkow, N. D. (2019). Brain default-mode network dysfunction in addiction. *NeuroImage, 200*, 313–331.

The Relationship Between Fear of Missing Out and Phubbing Behaviors: The Mediating Role of Addictive Smartphone Behaviors Among University Students

F. Muggianu, C. Sechi, C. G. Buyukbayraktar, and C. Cabras

Abstract This study aims to provide insight into potential risk factors for phubbing behavior, which has become increasingly prevalent in recent years despite limited research on this issue. A cross-sectional study was conducted with 599 Italian and Turkish university students (53.1% female) to investigate the mediating effect of Addictive Smartphone Behavior (ASB) on the relationship between Fear of Missing Out (FOMO) and Phubbing Behavior (PB). Participants completed surveys that included the Smartphone Addiction Scale Short Version (SASSV), the Fear of Missing Out scale (FoMOs), and the General Scale of Phubbing (GSP). Statistical analysis revealed significant direct and indirect effects among FOMO, ASB, and PB, with FOMO having a positive influence on both ASB and PB. Furthermore, there was an indirect influence of FOMO on PB via ASB. These findings contribute to a better understanding of the psychological variables underlying phubbing behavior.

Keywords Fear of missing out · Phubbing behavior · Addictive smartphone behavior · University students

F. Muggianu · C. Sechi · C. Cabras (✉)
Department of Pedagogy, Psychology, Philosophy, University of Cagliari, Cagliari, Italy
e-mail: ccabras@unica.it

F. Muggianu
e-mail: francesca.muggianu@unica.it

C. Sechi
e-mail: cristina.sechi@unica.it

C. G. Buyukbayraktar
Department of Social Work, Selcuk University, Beysehir-Konya, Turkey
URL: https://www.researchgate.net/profile/Cagla-Buyukbayraktar

G. Minati and M. Pietronilla Penna (eds.), *Multiple Systems*, Contributions to Management Science, https://doi.org/10.1007/978-3-031-44685-6_10

1 Introduction

Nowadays, the advent of smartphones has brought to several benefits. However, the ease of internet access and its broad availability have also resulted in undesirable outcomes. One such consequence is phubbing, occurring in a social situation, when an individual (the "phubber") disregards the conversation with an interlocutor (the "phubbee") by shifting the attention to the smartphone (Chotpitayasunondh & Douglas, 2016).

Several authors have identified unregulated internet use as a primary contributor to phubbing behavior (T'ng et al., 2018), and have also established a positive association between phubbing and addiction to smartphones (Chotpitayasunondh & Douglas, 2016). In recent years, Chotpitayasunondh and Douglas (2016) have identified FoMO, internet addiction, and self-control as the most significant predictors of phubbing. People who experience higher levels of FoMO are more susceptible to engaging in phubbing behavior due to their challenges in regulating social media and smartphone usage (Roberts & David, 2016). The implications of FoMO for addictive smartphone behavior have been highlighted in recent research (Buyukbayraktar, 2020), and its association with increased smartphone use is a major concern in modern society (Wolniewicz et al., 2018). Excessive mobile phone use can lead to negative consequences such as disruptions in daily routines and a distorted perception of reality (Bragazzi & Del Puente, 2014), underscoring the need for interventions to prevent and treat smartphone addiction.

Based on prior research, this investigation proposes the subsequent hypotheses: (1) Fear of Missing Out exhibits a positive correlation with Phubbing behaviors; (2) Fear of Missing Out demonstrates a positive correlation with Addictive smartphone behaviors; (3) Addictive smartphone behaviors are positively associated with Phubbing behaviors.

2 Material and Methods

The study adhered to the ethical guidelines set by the National Board of Italian Psychologists Code of Ethics for Psychologists. Non-probabilistic sampling was employed to recruit participants through advertisement messages posted on Italian and Turkish forums and social networks. Interested individuals who clicked on the link to the online survey were presented with an informed consent document outlining the survey's nature. The study included a total of 599 participants, with 318 females (53.1%) and 281 males (46.9%) ranging in age from 18 to 57 years, with a mean age of 24.7 years ($SD = 5.58$). Of the participants, 313 were Italian (52.3%) and 286 were Turkish (47.7%).

To measure fear of missing out, we used the Fear of missing out scale (FoMOs) developed by Przybylski et al. (2013). The scale includes 10 items rated on a five-point agreement scale (from 1 = not at all true for me, to 5 = extremely true of me).

In the present study, the FoMOs had high level of internal consistency ($\alpha = 0.83$ for Italians and $\alpha = 0.87$ for Turkish). To measure addictive smartphone behaviors, we used the Smartphone Addiction Scale Short Version (SASSV) developed by Kwon et al., (2013). The scale includes 10 items rated on a six-point agreement scale (from 1 = strongly disagree to 6 = strongly agree). In the present study, the SASSV had high level of internal consistency ($\alpha = 0.83$ for Italians and $\alpha = 0.89$ for Turkish). Phubbing behaviors were measured with General Scale of Phubbing (GSP) developed by Chotpitayasunondh and Douglas (2018). The scale contains 15 items rated on a seven-point Likert scale (from 1 = Never, 7: Always). The scale consists of four subscales: Nomophobia (NP), Interpersonal Conflict (IC), Self-Isolation (SI), and Problem Acknowledgment (PA). In the present study, the GSP had high level of internal consistency ($\alpha = 0.88$ for Italians and $\alpha = 0.93$ for Turkish).

3 Results

The hypothesized model was examined through structural equation modeling (SEM) using Amos 23.0. To evaluate the model fit, various fit indices were used, including chi-squared, CFI, TLI, SRMR, and RMSEA, along with 95% bootstrapped confidence intervals. Fit indices were deemed acceptable if the CFI and TLI were above 0.90, and the SRMR and RMSEA were below 0.08.

The results indicate that the hypothesized model fits the data well, as evidenced by the following fit indices: $\chi 2 = 33{,}658$, df = 8, $p = 0.00$, CFI = 0.98, TLI = 0.96, RMSEA = 0.07 (90% [CI]: 0.05–0.09), SRMR = 0.03. Specifically, Fear of Missing out had a direct impact on Addictive smartphone behaviors and Phubbing behaviors, supporting hypotheses H1 and H2. Furthermore, Fear of Missing out had an indirect impact on Phubbing behaviors through Addictive smartphone behaviors, supporting hypothesis H3. Finally, Addictive smartphone behaviors had a positive impact on Phubbing behaviors, supporting hypothesis H3.

4 Discussions

The objective of this research was to examine the relationships between fear of missing out (FoMO), smartphone addiction, and phubbing behavior. The results confirm previous research that has linked Fear of Missing Out (FOMO) with Phubbing behaviors (Chotpitayasunondh & Douglas, 2016). Individuals with high levels of FOMO often prioritize their social media and networking site usage over real-life interactions (Wolniewicz et al., 2018), leading them to inadvertently engage in phubbing behaviors (Vanden Abeele et al., 2016). Furthermore, our results demonstrate the potential risks associated with addictive smartphone behaviors and the development of Phubbing behavior. The constant availability and ease of access to smartphones and their various functions can contribute to addictive behaviors that negatively impact social interactions and lead to phubbing (Vanden Abeele et al., 2016).

Despite the societal challenges engendered by rapid technological advances, few empirical investigations have been undertaken to elucidate the psychosocial determinants of maladaptive technology use. Thus, the implications of this study are salient in illuminating the antecedents of phubbing behavior. By addressing these issues, this study endeavors to furnish valuable insights into effective interventions that can fortify social bonds in an increasingly disconnected world.

References

Bragazzi, N. L., & Del Puente, G. (2014). A proposal for including nomophobia in the new DSM-V. *Psychology Research and Behavior Management, 7*, 155.

Buyukbayraktar, C. G. (2020). Predictive relationships among smartphone addiction, fear of missing out and interaction anxiousness. *European Journal of Educational Sciences, 7*(2), 1–16.

Chotpitayasunondh, V., & Douglas, K. M. (2016). How "phubbing" becomes the norm: The antecedents and consequences of snubbing via smartphone. *Computers in Human Behavior, 63*, 9–18.

Chotpitayasunondh, V., & Douglas, K. M. (2018). Measuring phone snubbing behavior: Development and validation of the generic scale of phubbing (GSP) and the generic scale of being phubbed (GSBP). *Computers in Human Behavior, 88*, 5–17.

Kwon, M., Kim, D. J., Cho, H., & Yang, S. (2013). The smartphone addiction scale: development and validation of a short version for adolescents. *PloS One, 8*(12).

Przybylski, A. K., Murayama, K., DeHaan, C. R., & Gladwell, V. (2013). Motivational, emotional, and behavioral correlates of fear of missing out. *Computers in Human Behavior, 29*(4), 1841–1848.

Roberts, J. A., & David, M. E. (2016). My life has become a major distraction from my cell phone: Partner phubbing and relationship satisfaction among romantic partners. *Computers in Human Behavior, 54*, 134–141.

T'ng, S. T., Ho, K. H., & Low, S. K. (2018). Are you "phubbing" me? The determinants of phubbing behavior and assessment of measurement invariance across sex differences. *International and Multidisciplinary Journal of Social Sciences, 7*(2), 159–190.

Vanden Abeele, M. M., Antheunis, M. L., & Schouten, A. P. (2016). The effect of mobile messaging during a conversation on impression formation and interaction quality. *Computers in Human Behavior, 62*(C), 562–569.

Wolniewicz, C. A., Tiamiyu, M. F., Weeks, J. W., & Elhai, J. D. (2018). Problematic smartphone use and relations with negative affect, fear of missing out, and fear of negative and positive evaluation. *Psychiatry Research, 262*, 618–623.

Symbolic Processing as the Result of Social Interactions

Simone Pinna, Fabrizia Giulia Garavaglia, and Marco Giunti

Abstract According to Edwin Hutchins (in Cogn Sci 19:265–288, 1995a; Cognition in the Wild. MIT Press, 1995b), cognitive science should explain how humans became skilled processors of symbolic structures, rather than assuming that the architecture of cognition is symbolic. He suggests that the origin of symbolic processing should be traced back to human interaction in sociocultural systems. Developing Hutchins' ideas, the distributed approach to language grounds linguistic abilities in proto-symbolic exchanges that can be observed in early child–parent interactions (Spurrett and Cowley in The Extended Mind. The MIT Press, 2010). Here, the parent's adoption of a *language stance* (i.e., the assumption that children's gestures and utterances have a symbolic content) leads to the reinforcement of the child's linguistic behavior, which in turn leads the child to adopt a linguistic stance toward other people, and so on. In this view, the parent–child pair forms a *dual system* from which the language stance emerges. In order to envisage an explanation to human's symbolic processing abilities in a distributed perspective, we propose to add a further step to this hypothesis. Once language is adopted as the privileged vehicle for interaction, a *symbolic stance*, in which symbols are understood *as if* they were well-codified objects of a specific language, emerges on a *multiple system* composed of (at least) a cognitive subject, the social community of users of certain symbolic systems, and the conventions and norms associated with the use of these symbolic systems.

S. Pinna (✉) · F. G. Garavaglia · M. Giunti
Università Di Cagliari, Cagliari, Italy
e-mail: simonepinna@hotmail.it

F. G. Garavaglia
e-mail: giuliafabrizia@gmail.com

M. Giunti
e-mail: giunti@unica.it

G. Minati and M. Pietronilla Penna (eds.), *Multiple Systems*, Contributions to Management Science, https://doi.org/10.1007/978-3-031-44685-6_11

1 Introduction

Distributed cognition assumes the general hypothesis that all cognition emerges from the interaction among elements in distributed systems (Hutchins, 2014). In this paper, we address the problem suggested by Hutchins (1995b) of explaining how humans learn to be skilled processors of symbolic structure in the context of cultural socio-systems.

First, we provide a general characterization of distributed cognition.

Second, we present a hypothesis of language development in the context of distributed cognition (Spurrett & Cowley, 2010). This hypothesis predicts a continuous line of development from proto-linguistic behavior to the consistent use of linguistic symbols through the notion of a *language stance*, which emerges from a parent–child *dual system* and allows the use of linguistic modalities as the privileged vehicle for interacting with others.

However, this hypothesis alone cannot fully explain how we acquire the use of well-defined and culturally inherited *symbolic systems*. As a further step, we propose that the use of language as a privileged vehicle for interaction, made possible by the emergence of a language stance, triggers a complex set of cognitive skills that, eventually, lead to the adoption of a *symbolic stance*, in which symbols are treated *as if* they were objects that refer to a specific meaning and set of rules determined by the sociocultural context in which they are used. The emergence of a symbolic stance is characterized by reference to a broader sociocultural context than the dual system that governs the early stage of linguistic development. In fact, it emerges on a *multiple system* consisting of (at least) a cognitive subject, the social community of users of a given symbolic system, and the conventions and norms that all users should follow in order to make consistent use of that symbolic system.

Finally, we show that our hypothesis fits with an explanation of how children acquire basic knowledge of the natural number system, which is the first non-linguistic symbolic system that we learn to use from infancy.

2 The Distributed Perspective

Hutchins (1995a, 1995b) analyzes the processes that take place inside the cockpit of an airliner as a typical example of distributed cognition. In this particular case, the cognitive work involves two people (the pilots) and the cockpit instruments they use following meticulously predetermined routines in order to avoid any kind of pilot error. Hutchins' analysis highlights two distinctive features of this distributed system. First, the cockpit instruments play a real cognitive role, in that many of the parameters relevant to the performance of crucial maneuvers are left as *distributed representations* that make up the *cockpit memory* (e.g., the moving notches called *speed bugs* set on the speed indicator that are used to change the wing configuration during a landing maneuver). Second, the cognitive work performed by the distributed system (pilots + cockpit) takes the form of a highly prearranged *cultural practice*, as it follows routines, conventions, and socially codified norms that have been developed on the basis of past experience to deal with problems that might arise in this particular type of work.

According to Hutchins, the processes that can be observed in such a system are cognitive in Simon's (1981) classical sense because they involve

> how representations are transformed, combined, and propagated through the system [...] With the new unit of analysis, many of the representations can be observed directly, so in some respects, this may be a much easier task than trying to determine the processes internal to the individual that account for the individual's behavior. Posing questions in this way reveals how systems that are larger than an individual may have cognitive properties in their own right that cannot be reduced to the cognitive properties of individual persons. (Hutchins, 1995a, p. 266)

In Hutchins' view, however, cognition is not constitutively based on symbolic representations and transformations, as, for instance, in Newell's physical symbol system theory (Newell, 1980). In contrast, Hutchins argues that human competence as symbolic processors is not primitive, but is derived from cognitive strategies used in sociocultural contexts:

> There are no plausible biological or developmental stories telling how the architecture of cognition became symbolic. We must distinguish between the proposition that the architecture of cognition is symbolic and the proposition that humans are processors of symbolic structures. The latter is indisputable, the former is not. I would like to be able to show how we got to be symbol manipulators in relation to how we work as participants in sociocultural systems, rather than assume it as an act of faith. (Hutchins, 1995b, p. 369)

For Hutchins, then, the distributed approach allows for a general, empiricist theory of the roots of cognition, whereas the symbolic approach often relies on nativist explanations to justify the observed cognitive abilities.[1]

The case of distributed cognition described above is obviously consistent with extended cognition (Clark, 1997, 2008). Indeed, insofar as a cognitive strategy implies the use of any kind of external resources with respect to the brain and/or body of a cognitive subject, this strategy can be described in terms of extended cognition. It would seem, therefore, that distributed and extended cognition is indeed a unique approach, but this is not the case. Hutchins (2014) clarifies this issue by pointing out that distributed cognition is not a *kind* of cognition, but a *perspective on all of cognition.* In fact, extended cognition is an empirical hypothesis according to which *some* cognitive systems can be seen as distributed between the cognitive subject and its surrounding space. In Hutchins' view, extended cognitive systems are special cases of distributed systems «in a specific range of spatial and temporal scales» (Hutchins, 2014, p. 35). Distributed cognition, on the other side, assumes only the general view that *all cognition emerges from the interactions among elements in distributed systems*:

[1] A famous example of this kind of explanations, perfectly consistent with the symbolic approach, is Noam Chomsky's theory of language (Chomsky 1965, 1967; Hauser et al., 2002).

> The hypothesis of extended cognition is an important hypothesis within the perspective of distributed cognition. Such hypotheses are subject to confirmation or rejection via empirical investigations. However, there is no series of experiments or set of observations that could prove or disprove the distributed cognition perspective, because the perspective itself makes no empirical claims. (Hutchins, 2014, pp. 36–37)

Unlike Clark's version of extended cognition (Clark, 2008), distributed cognition does not pose a center of cognition, i.e., a core or most active element for all cognitive phenomena.[2] Moreover, it does not refer only to *ecological assemblies* at intermediate spatiotemporal scales, but also to *cultural ecosystems* operating at larger spatial and temporal scales.

To sum up, the distributed perspective includes extended cognition as one of its cases that can be observed at a mid-level scale (ecological assemblies), but, in addition, it enables the study of cognitive system at a low-level scale (such as the brain, whose processes can quite obviously be described from a distributed perspective), and, notably, large-scale phenomena that, in general, concern the development and evolution of cognitive abilities as the result of sociocultural practices that can be observed in specific cultural ecosystems. Distributed cognition, therefore, undertakes the task of describing and studying such cultural ecosystems.

In the following section, we present an interesting hypothesis, within the distributed perspective, on language development. We then extend this hypothesis in order to address Hutchins' question about the capabilities of humans as symbol processors.

3 Language Stance and Symbolic Stance

One of the goals of distributed cognition is to provide a common framework for explaining cognition, action, and perception. Stephen Cowley, the most vocal proponent of the distributed approach to language, argues that a key concept for understanding *languaging* (a term for the connection between cognition, action, and perception relative to language use) is that of *utterance activity*:

> "Utterance-activity" is a term of art [. . .] used here to refer to the full range of kinetic, vocal, and prosodic features of the behavior of interacting humans. Utterance-activity can include, but is not restricted to, what are usually regarded as words and strings of words. [. . .] Further, we regard it as in crucial ways continuous with, and inextricable from, (nonwritten) language. (Spurrett & Cowley, 2010, pp. 295–296)

The concept of utterance activity aims to highlight the continuity between language and other expressions of human behavior. In contrast to the traditional accounts of linguistics (in the line developing from de Saussurre to Chomsky),[3] whose primary object of explanation is language understood as a fixed code, the

---

[2] According to Clark's *hypothesis of organism-centered cognition*, the core of cognition is (obviously) the brain (Clark 2008, p. 39).

[3] See, for example, de Saussurre (1922), Chomsky (1965, 1967).

distributed view of language focuses on specific properties of spoken language (such as gesture, pronunciation, prosody, etc.). According to Cowley, utterance activity is a key concept for understanding the development of language skills from more basic forms of behavior. Indeed, the analysis of linguistic interactions can provide many interesting insights into the explanation of language evolution, both from an onto- and phylogenetic[4] perspective. Moreover, given the centrality of language abilities in human cognition, this explanation returns a picture of cognition as fundamentally intertwined with social interaction.

According to this approach, language abilities should not be considered as properly individual abilities. When we analyze a conversation, we must consider a distributed system that includes the subjects participating in the conversation and the linguistic community to which they refer. This cognitive system extends into the proximal (physical) space around the subjects and into the conceptual space formed by their linguistic community. Moreover, this system is also extended in time, that is, it includes references to the past time in which all the beliefs, traditions, behavioral norms, etc. of that community were formed (Cowley, 2011). Similar considerations are useful to illustrate situations where we use language in non-colloquial contexts. For example, when we write a text, our actions can be adequately explained as the result of a system that includes ourselves, the linguistic community to which the text refers, and the physical object constituted by the text itself. Interaction with a written text, in fact, should not be seen exclusively in terms of an actual externalization in written form of the contents encoded in our minds; rather, through a continuous interplay of feed-back, written words (like any kind of external symbol) are able to influence our thinking and contribute to the formation of higher-order cognitions.

3.1 Linguistic Development

Within this framework, Spurrett and Cowley (2010) propose a precise hypothesis about the onset of language behavior in children. Their explanation is based on a series of observations, in an anthropological context, of actual linguistic (or rather, proto-linguistic) interactions between mother and child (Cowley et al., 2004). In these cases, some episodes of "overinterpretation" are revealing of what is going on:

> A further episode from our data, in this case concerning a child of around four months, illustrates this point about overinterpretation. In it an infant repeatedly vocalizes in ways which to its mother, although not to us, are suggestive of its saying "up." After several responses along the lines of "up?" or "you want to go up?" the mother lifts the child. Prior to this, to the detached observer, there is little evidence that the child actually wants to be lifted, or that its attention is focused on anything at all. When it is lifted, though, it beams

[4] A hypothesis on the origin of language from lower-level cultural practices has recently been proposed in Cowley and Kuhle (2020). Here, the continuity between animal and human cognition is made visible through a comparison between socially/culturally learned practices in higher primates (in particular, the use of instruments to gain food) and languaging in non-literate societies.

widely. Whatever it did want, if anything, it is now, we suggest, one step closer to figuring out how to behave in ways that lead to its being lifted up. (Spurrett & Cowley, 2010, p. 307)

When the mother in the story starts to interpret her child's vocalizations, she is taking a *language stance* (Cowley, 2011). This expression, modeled on Dennett's *intentional stance*, indicates the projection on others of our linguistic understanding capacities. According to Dennett, the level of intentionality that we are inclined to attribute to creatures different from us (be them other human beings, animals, robots, etc.) is largely dependent on the projection of our mental contents on them, namely, we treat others as if they were intentional agents. This enables us to predict others' behavior and interact with them (Dennett, 1989, 1991). For example, if a dog in front of me wags its tail or whines, I will be predisposed to behave toward it *as if* it were happy or sad, even without knowing at all how a dog feels happiness or sadness, that is, *without knowing anything about what it feels like to be a dog*. Linguistic stance, similarly, consists of projecting onto other subjects the capacities of linguistic understanding that belong to us. This, in turn, allows cases of *overinterpretation*, as in the above example, in which gestures and vocalizations are taken as genuine linguistic behaviors.

The hypothesis of linguistic development in this approach may be summarized as follows: initially, the interaction between parent and child—mediated by proto-linguistic modalities—enables the parent's adoption of a language stance with respect to the child's behavior; this process acts as a reinforcer for the child's linguistic behavior, which in turn is induced to take a language stance and, as a consequence, to use linguistic modalities as a privileged vehicle for interacting with others, and so on. The adoption of a linguistic stance, in other words, triggers a virtuous circle in which the child gradually becomes aware of the relationship between his or her behavior and the parent's response, while the parent becomes increasingly confident in the child's ability to express meanings through his or her linguistic behavior. The aspects of utterance activity that characterize this early stage of language development remain central in adulthood, as evidenced by the importance of prosodic properties of speech and *vocal gestures* for communicating semantic nuances (Spurrett & Cowley, 2010, Sect. 3).

This hypothesis is meant to be a way to answer the crucial question: «*How can anything come to count as a symbol?*» (Spurrett & Cowley, 2010, p. 300). Indeed, the child's behavior in the above example is understood by the mother *as if* it was a way to communicate the child's desire to be lifted, so it represents a symbol of this desire. In order to understand this point, we need to be more specific about the sense in which the word symbol is being used in this approach. This word does not refer to a term that is given a conventional meaning within a formal system as understood in classical cognitive science. Rather, the concept of symbol in this sense is similar to that of *sign* according to Harris (2000). In opposition to the classic idea of language as a fixed code, Harris proposes an *integrational* view of language, where language is seen as a *second level cultural construction* based on the (first level) activity of producing and interpreting *linguistic signs*.[5] This activity, in particular, consists of

[5] See also Love (2004).

attributing (in real time and in highly context-sensitive situations) semiotic meaning to different behaviors (vocal, gestural, etc.) which are, in this process, integrated:

> Signs, for the integrationist, provide an interface between different human activities, sometimes between a variety of activities simultaneously. They play a constant and essential role in integrating human behaviour of all kinds [...] Signs are not given in advance, but are made. The capacity for making signs, as and when required, is a natural human ability. (Harris, 2000, p. 69, quoted in Love, 2004, p. 531)

A sign, in this view, is not a definite object like the linguistic units of some fixed code, as in the traditional accounts of linguistics, but is highly context dependent. For example, a road sign has the precipitous role of a sign not in itself, but only to the extent that someone interprets it that way. Moreover, the interpretation need not be unambiguous. The sign indicating a speed limit will mean to a bus driver that you are not allowed to exceed a certain speed on that stretch of road and that if you exceed it, you may incur a fine. For a passenger on the same bus, however, the same sign may have a different meaning, for example of relative proximity to a certain stop, so once you reach that sign you should approach the exit door. As can be seen from this simple example, it is the context that determines the meaning of a sign, so any idealization that obfuscates what characterizes the concrete use of linguistic signs will necessarily return a misleading picture of language.

3.2 *Symbolic Stance*

The integrationists' idea of language broadens the range of phenomena that can be understood as linguistic. In fact, it is not possible to identify language with verbal behavior alone because, for example, an appropriate but non-verbal response to a verbal request (such as handing the saltshaker to someone who asks, "would you hand me the salt?") still represents a manifestation of linguistic knowledge.

However, the distinction between the first-level activity of producing language symbols (in the broadest sense) and the second-level activity of using symbols of a specific language with all its culturally formed conventions and norms is seen in a continuous path, in which the abilities acquired during the first-level activity are a necessary condition for the use of any specific language. In this sense, the language development hypothesis seen (Sect. 3.1) should be viewed as a first step toward explaining how humans acquire symbol manipulation skills in the context of sociocultural systems (Hutchins, 1995b, p. 369). To this end, we propose a further step in the hypothesis. We have seen that, to explain how humans learn symbol use, Spurrett and Cowley (2010) propose that this ability emerges from the activity of a parent–child *dual system* that, in its interaction, eventually adopts a language stance at both poles. As a further step, we propose that the use of language as a privileged vehicle of interaction, made possible by the assumption of a language stance, triggers a complex set of cognitive skills that, eventually, lead to the adoption of a *symbolic stance*, in which symbols are not *only* seen in the broader sense of highly context-dependent semantic vectors but also as precise objects of a specific language with all its culturally formed conventions and norms that, in general, involve a more or less clearly defined syntax and semantics.

In adopting a symbolic stance, then, symbols are treated *as if they were objects that refer to a specific meaning and set of rules decided by the socio-cultural context in which they are used.* The emergence of a symbolic stance is characterized by reference to a broader sociocultural context than the coupled system that governs the early stage of language development. In fact, it emerges on a *multiple system* consisting of (at least) a cognitive subject, the social community of users of given symbolic systems, and the conventions and norms that all users should follow in order to make consistent use of that symbolic system.

4 Acquiring Basic Numerical Knowledge

As specified in the previous section, assuming a symbolic stance is a preliminary condition for referring symbolic objects to specific meanings and sets of transformation rules. In other words, once we have adopted a symbolic stance, we are in a position to attribute some of the symbols that we use to particular *symbolic systems.* If this is true, then we can trace a continuous line of development from the early stages of proto-language to the typically human high-level symbol-processing skills, thus envisaging an explanation of the latter cognitive abilities by a distributed perspective.

The first non-linguistic symbolic system we naturally encounter from childhood is the natural number system. Children acquire basic numerical knowledge through counting skills (Gelman & Gallistel, 1978). First, they learn number words, i.e., an ordered list of words (numerals) with no numerical meaning. This list is then used for counting objects. A classic counting situation can be described as follows: A group of objects is placed in front of a child who points to them one at a time, assigning to each one a precise numerical label from the ordered list. The skill is acquired when the child learns that the label assigned to the last object counted represents the cardinal number of the set of objects. It is worth noting that, at this point, children have actually acquired the knowledge of at least two *algorithms* for computing the corresponding functions defined on a finite subset of natural numbers: the function that assigns to each number word the corresponding cardinal number, and the *next number*, namely the successor function. This requires knowledge of a simple well-defined symbol structure (see Giunti & Pinna, 2016).

A central issue in numerical cognition[6] is the individuation of the basic cognitive resources that enable knowledge of the numerical system. Researchers agree that the neurophysiological basis of this knowledge lies in brain modules dedicated to the representation of numerical quantities. However, this low-level explanation cannot account for the influence of social practices in the acquisition of numerical knowledge. Indeed, it has been proposed that number processing is influenced by culture, for example, by finger-counting habits (Bender & Beller, 2012; Domahs et al., 2010).

[6] The main issues in this topic are discussed in Dehaene and Brannon (2011).

The distributed perspective is not affected by this (apparent) contrast between low-level and higher-level explanations. Indeed, the neural representation of numerical quantities can be seen as a necessary condition for the acquisition of numerical knowledge, but it is not sufficient without the emergence of a symbolic stance on a multiple system of which the cognitive subject, with its neural structure, is only one component. The symbolic stance, then, links the neural representations of quantities to the social practices and conventions that give the symbolic system of natural numbers its structure. Moreover, our proposal is neutral with respect to the debate on empiricism vs conventionalism in numerical knowledge.[7] In fact, both empirical (neural) and conventional (cultural) components are necessary for the emergence of a symbolic stance.

In summary, in order to provide a suggestion for an explanation of human symbolic abilities in a distributed perspective, we propose to add one more step to Spurrett and Cowley's (2010) hypothesis on the development of linguistic abilities. Indeed, the adoption of a language stance, in which symbols are understood in all their complexity as linguistic signs, is seen as a prelude to the adoption of a symbolic stance, in which the use of symbols is governed by conventions and norms developed in a given sociocultural context.

References

Bender, A., & Beller, S. (2012). Nature and culture of finger counting: Diversity and representational effects of an embodied cognitive tool. *Cognition, 124*(2), 156–182.
Chomsky, N. (1965). *Aspects of the theory of syntax*. MIT Press.
Chomsky, N. (1967). Recent contributions to the theory of innate ideas. *Synthese, 17*, 2–11.
Clark, A. (2008). *Supersizing the mind. Embodiment, action and cognitive extension, Oxford*. Oxford University Press.
Clark, A. (1997). *Being there: Putting brain*. The MIT Press.
Cowley, S. J. (2011). Taking a language stance. *Ecological Psychology, 23*, 185–209.
Cowley, S. J., & Kuhle, A. (2020). The rise of languaging. *Bio Systems, 198*, 104264.
Cowley, S. J., Moodley, S., & Fiori-Cowley, A. (2004). Grounding signs of culture: Primary intersubjectivity in social semiosis. *Mind, Culture, and Activity, 11*, 109–132.
de Saussure, F. (1922). *Cours de linguistique générale*. Payot.
Dehaene S., & Brannon E. (Eds.). (2011). *Space, time and number in the brain: Searching for the foundations of mathematical thought*. Academic Press.
Dennett, D. C. (1991). *Consciousness explained*. Little Brown and Company.
Dennett, D. C. (1989). *The intentional stance*. MIT Press.
Domahs, F., Moeller, K., Huber, S., Willmes, K., & Nuerk, H.-C. (2010). Embodied numerosity: Implicit hand-based representations influence symbolic number processing across cultures. *Cognition, 116*, 251–266.
Gelman, R., & Gallistel, C. (1978). *The child's understanding of number*. Harvard University Press.
Giunti, M., & Pinna, S. (2016). For a dynamical approach to human computation. *Logic Journal of the IGPL, 24*(4), 557–569.
Harris, R. (2000). *Rethinking writing*. Athlone Press.

[7] For an overview and an interesting solution of this debate, see Pantsar (2014).

Hauser, M. D., Chomsky, N., & Fitch, W. T. (2002). The faculty of language: What is it, who has it, and how did it evolve? *Science, 298*, 1569–1579.
Hutchins, E. (1995a). How a cockpit remembers its speeds. *Cognitive Science, 19*, 265–288.
Hutchins, E. (1995b). *Cognition in the wild*. MIT Press.
Hutchins, E. (2014). The cultural ecosystem of human cognition. *Philosophical Psychology, 27*(1), 34–49.
Love, N. (2004). Cognition and the language myth. *Language Sciences, 26*, 525–544.
Newell, A. (1980). Physical symbol systems. *Cognitive Science, 4*, 135–183.
Pantsar, M. (2014). An empirically feasible approach to the epistemology of arithmetic. *Synthese, 191*(17), 4201–4229.
Simon, H. A. (1981). *The sciences of the artificial* (2nd ed.). The MIT Press.
Spurrett, D., & Cowley, S. J. (2010). The extended infant: Utterance-activity and distributed cognition. In R. Menary (Ed.), *The extended mind* (pp. 295–323). The MIT Press.

Coevolution Dynamics and the Biosemiotics of Human Change

Franco F. Orsucci

Abstract *Scope* Clarify how coevolution, cooperation, and creation dynamics are crucial in human change. *Method* Exploring the paradox of private language, synchronization theory, complex biosemiotics, and heterogeneous dynamics involved in therapeutic relationships as a case study for general human change dynamics. We examined specific theoretical implications of our research groups' empirical studies developed over recent years. We explored the biosemiotic nature of communication streams from emotional neuroscience and embodied mind perspectives. They investigated intraindividual and interpersonal relations and found coevolution dynamics based on hybrid couplings, synchronizations, and desynchronizations. Cluster analysis and Markov chains produced evidence of chimera states and phase transitions. A probabilistic and non-deterministic approach clarified the properties of these hybrid dynamics. As a result, multidimensional theoretical models can better represent the hybrid nature of human interactions. We presented and discussed different models. *Results* New challenges lie ahead. Complex systems present plural structural forms and varieties of organization and disorganization. These varieties are frequently distributed within any singular system, creating rugged dynamical landscapes. Hybrid forms of organization are specific to human systems, as they present multiple scales and heterogeneous subsystems; synchronous and asynchronous interactions; stable, unstable, and metastable states; and localized and generalized dynamics. Therefore, considering the hyper complexity of the human dynamical landscapes, empirical studies can use mixed methods and different approaches (sometimes all at once). Accordingly, multiple and varied interventions can induce change or facilitate its natural evolution and emersion. Therefore, multiple and different therapeutic techniques can produce similar (though not identical) outcomes in the clinical field. The good old equifinality principle of complex open systems is still at work with more advanced science tools. Coevolutionary cooperative creations present robust evidence in human change dynamics. We finally examine the personalization and ethical aspects of these models.

F. F. Orsucci (✉)
University College London, London, England
e-mail: f.orsucci@ucl.ac.uk

Norfolk and Suffolk NHS Foundation Trust, Norwich, United Kingdom

G. Minati and M. Pietronilla Penna (eds.), *Multiple Systems*, Contributions to Management Science, https://doi.org/10.1007/978-3-031-44685-6_12

1 Introduction. A Paradox on Solipsism and Cooperation

As a form of thought experiment, we might consider the so-called private language argument (Orsucci, 2015). In some well-known paragraphs of the Philosophical Investigations, Ludwig Wittgenstein posed this problem: "The words of this language relate to what can be known only to the speaker, to his immediate and private sensations. So, no one else can understand this language." §243 described: "The words of this language refer to what only the speaker can know — to his immediate private sensations. So, another person cannot understand the language." (Wittgenstein, 1967). Wittgenstein considered a language intelligible only to its single originator and inaccessible to others.

As an example of private language, we might consider the Voynich Manuscript, an illustrated codex hand-written in an unknown, possibly meaningless, writing system (Clemens, 2016). Wilfrid Voynich, a Polish book dealer, purchased the manuscript in 1912. The material has been carbon-dated to the early fifteenth century (1404–1438), possibly composed in Italy during the Renaissance. The Voynich Manuscript has been studied by many professional and amateur codebreakers but never deciphered. Nevertheless, it is fascinating that the manuscript became a *cause celébre* in cryptology and, at some point, even a subject for novels.

Nevertheless, the mystery of its meaning and origin has excited the popular imagination, making it the subject of study and speculation. In 1969, Hans P. Kraus donated the Voynich Manuscript to Yale University's Beinecke Rare Book and Manuscript Library, where it remains. Additional resources about it are available on the library website http://www.voynich.nu. The Voynich Manuscript, though written in unusual signs, presents an apparent natural language structure in sequences and recurrences (Barlow, 1986; Brumbaugh, 1975, 1978; Landini, 2001). We find an imaginary linguistic universe with unusual signs, grammar, syntax, and a dictionary.

Would it then be possible to conceive a peculiar language that is not understandable to anyone except its creator, who might use it just for private writing, thoughts, and solitary conversations? After introducing the idea, Wittgenstein clarifies that, from his point of view, there can be no private language because all language is essentially the result of public interpersonal and social interactions, which he calls "language games." The importance of drawing philosophers' attention to an unusual idea and then arguing that it is impossible lies in the fact that implicit confidence in the possibility of a private language has been essential to idealistic epistemology and metaphysics. Most representational theories of mind imply some privacy of mental contents. This position is significant in philosophy, linguistics, psychology, and epistemology (Baker, 1998; Kripke, 1982). Wittgenstein's paradox is relevant because the possibility of a private language is a vague assumption of standard theories of knowledge and philosophy of mind from Plato to Descartes and most of the cognitive science of the late twentieth century.

Wittgenstein dedicated his investigations to "healing grammar diseases" and "infections" from conceptual confusion. Similarly, early Freud's attention went to all the anomalies that jut out of speech—puns, slips of the tongue, and narrative of dreams, regarded as rebuses or cryptic texts needing deciphering. For Wittgenstein, language is not intrinsically misleading or deviant. On the contrary, Wittgenstein seems to find the cause of "linguistic disorders" in a metaphysical desire to "[run] its head up against the limits of language" (Wittgenstein, 1967). So, in principle, he confirms that grey regions of uncertain language privacy might be possible. However, from his point of view, they might represent forms of the linguistic disorder. From that point of view, thinking and language disorders have developed over the centuries spreading like epidemics. Wittgenstein would rank idealism and metaphysics between these linguistic syndromes at the origins of socio-political catastrophes and therapeutic disasters. In these dynamics of linguistic epidemiology, the meme construct is a source for further reflection and research (Blackmore, 2000; Dawkins, 2006). The private language paradox could find its roots in the ambiguous phase transitions between language in public communication and thinking in internal dialogue. Human language has the remarkable property of admitting usage for interpersonal communication and thinking, often in inner monologues or virtual dialogues. Not surprisingly, a line of linguistic research has suggested the notion of idiolect as a form of unique individual semiotic identity pattern (George, 1990; Kraljic et al., 2008). The idiolect manifests itself in lexicon forms, idioms, grammar, and pronunciation unique to us (Higginbotham, 2006).

Many states of mind can alter semantic fields and individual syntax and grammar changes (Lacan, 1978; Lorenzer, 1977; Merleau-Ponty, 1960; Orsucci, 1981a, 1981b). It seems that mental conditions correlate with changes in thinking and social communication. Every psychosocial, cultural, and therapeutic practice occurs in the grey areas between private dialogue and public conversation. Communication is constantly developing in the transitional/transactional areas of an as-if understanding. The psychopathology of language highlights how neologisms and neo-languages can emerge in mental health conditions (Andreasen & Grove, 1986; Condray et al., 2002; Crow, 2000; DeLisi, 2001).

Language develops in a coordinated flow of interactions, not in single gestures, sounds, words, or attitudes extracted from the communication streams. Inner and outer streams constantly merge or diverge. Language is a manner of living together in a flow of consensual behaviors (Maturana & Varela, 1980). Language evolves as a form of structural coupling in which living systems interact, engage, and affect each other. The concept of coupling between systems is present both in physics and biology. Coupling can facilitate the entrainment in synchronization (Pikovsky et al., 2001) and coevolution (Durham, 1992; Lumsden et al., 1981). In this perspective, following the pathways traced by Wittgenstein (1967), language can be considered a tool for cognition and social interaction. Language can produce the weaving of presentations between agents in distributed cognitive systems. Language is "a skillful, joint activity through which interlocutors attune to each other and the task at hand co-constructing a shared cognitive niche" (Fusaroli & Tylén, 2016; Fusaroli et al., 2014). A language is a form of joint action through multiple attunement/coupling dynamics of cognitive and motor processes (Tschacher & Bergomi, 2015).

The standard approaches to social cognition rarely emphasize intersubjective interaction, and even when they mention interaction, they frame the problem in terms of two minds communicating across a gap (Gallagher, 2001, 2013). In the theory of mind (ToM), inference about what is going on in others' minds bridges the grey area gap between minds. The ToM bridging inference is a simulation that will permit a form of mind-reading, also called "mentalizing." However, we do not usually take that detached observational stance in everyday social life with others. As agents and partners of the situation, we directly interact with our own verbal, motor, and emotional systems. Developmental studies prove this active process that the developmental psychologist Colwyn Trevarthen calls primary intersubjectivity (Trevarthen, 1998). By the end of the first year of life, the infant develops a non-mentalistic, perceptually based, embodied understanding of the intentions and dispositions of another person. These capabilities mature and become more sophisticated (see Dittrich et al., 1996), as shown in a micro-analysis of people's postures, movements, gestures, gazes, and facial expressions as they engage in a novel task. Communication among them is intrinsic to the actions that they take. This primary intersubjectivity is supplemented and enhanced by secondary intersubjectivity (Trevarthen, 2012). We start later to notice and critically reflect on how others interact with the world in expressions, intonations, gestures, and movements, and the bodies manifest in different contexts. As Gibson's theory of affordances (Gibson, 1979) suggests, we see things in our interactions and never as disembodied observers.

Dynamical system theory can be an excellent toolbox for understanding the synergy between symbolic and dynamic aspects of language (Elman, 1996; McWhinney, 1999; Orsucci, 2002; Rączaszek-Leonardi & Kelso, 2008). Human language evolved—through a series of intermediate stages of mimetic culture—due to pressure for increasingly sophisticated means of socio-cultural coordination and cooperation (Orsucci, 2008a, 2008b). The main features deriving from the dynamical system nature of language are: (1) higher balance of dynamical synchronization between interlocutors, increased stability, and complexity of collectively evolved symbolic patterns, (2) better coordination and, therefore, higher synergy performance in everyday tasks, and (3) emerging symbolic patterns of trans-subjective cultural repositories of collective knowledge. Semiotic patterns come to constitute communal cognitive niches structuring and coevolving with coordinative dynamics. Semiotics can be considered a tool for social and ecological harmonization. Synchronization in complex heterogeneous networks produces the emergence of semantic patterns from indexes, icons, and symbols (Orsucci et al., 2013, 2015, 2016). These studies of complex intersubjective dynamics are the evidence-based alternative to the idealism of monologue-based monadic rationalism—the same rationalism of the cognitivist approaches to language, such as Generative grammar.

2 Materials and Methods. Attachment, Love, and Cooperation

Coevolving living systems can establish different types of relationships relevant to ecology and social sciences, including psychology and the specialized forms of human relationships, also called psychotherapy (Holstein & Gubrium, 2012; Margulis et al., 2011; Orsucci, 2002, 2013). A relevant feature of coupling between human systems is that symmetry makes them different from other asymmetric couplings in empirical research. Freud recognized this specific order parameter when *transference* mirrors *countertransference* (Laplanche & Pontalis, 1974). The therapist resonates with the patient in a *quasi-symmetric coupling and coevolution.*

Coevolution can take different routes while developing in a system open to a broader environment of ecosystems. Coevolution between different species can evolve in commensalism, parasitism, competition, or mutualism dynamics. *Commensalism* represents a class of relationships between two organisms where one benefits but is unaffected. It is perhaps the most unstable and challenging to recognize because it can easily slip into other types of relationships. *Parasitism* is a symbiotic relationship where one organism, the parasite, benefits at the host's expense. As a result, the parasite ends up hurting the vital capacity of the host for its benefit. *Competition* can be *cooperative or predatory. Cooperative competition* promotes the vital capacity of competitors, and it is a challenge that strengthens both parties. This kind of competition tends to fade quickly in mutualism. *Predatory competition* is a coevolutionary process of damaging or eliminating the rival system.

In some cases, it looks like a *zero-sum game* where the winner takes all because the success of one of the competitors implies eliminating the other. However, total elimination of prey could starve predators, so more frequently, some coevolutionary balance settles before the zero-sum result. For example, *mutualism* is a *cooperative coevolution* in which each participant benefits from the interaction. Humans present this form of coevolution as the foundation of communities and the wider society.

Some questions are still partially open about cooperative dynamics. For example, why should an individual carry out a cooperative behavior that appears costly to perform but benefits others (Hamilton, 1964)? Benefits could be classified as direct or indirect fitness if linked to direct or indirect genetic benefits. However, these are not mutually exclusive, as, at times, they could be complementary (West et al., 2007a, 2007b). Nowak (2006, 2013) proposes five mechanisms for the evolution of cooperation: kin selection, direct reciprocity, indirect reciprocity, network reciprocity, and group selection. He highlighted how having a theory that can only explain cooperation among relatives is unsatisfactory. We also observe cooperation between unrelated individuals or members of different species. There are other forms of human payoff beyond genetics—societal factors such as reputation, influence, and other material and immaterial benefits. Ethical and spiritual factors also have a relevant function in social welfare and cohesion, affecting the fitness of associated individuals. On some rare occasions, the therapeutic pathway can deviate from the dynamics of mutualism and cooperation to other potentially unhealthy domains. The clinical community recognized the risks of dynamic drifts. Searles (1979), for example, discussed the paradox of situations in which patients could take inverted roles, behaving as if therapeutic for the therapist.

Relationships of care should follow *mutualism, cooperation* that develops through structural coupling between partners. The therapist drives the process by perturbing and accepting to be perturbed in controlled ways, facilitating opportunities for change through new self-organization. Both partners in coevolution will change during this process. The structural coupling in psychotherapy creates what some authors define as an *inter-subjective bi-personal field* (Baranger & Baranger, 1985; Langs, 1976). This field includes *multiple semiotic, motor, and emotional variables* generated through various small reciprocal perturbations. It is a process of gentle "pushing and pulling," using a metaphor from the physics of coupling and synchronization (Carroll & Pecora, 1993; Dube & Despres, 2000; Pyragas, 2001; Strogatz, 2003). Alternate phases of coupling and decoupling punctuate the dynamic change process (Heagy et al., 1995). The therapist intends to drive a transformative process through sequences of mutual perturbations and waves of emotional and cognitive interaction (Guidano, 1991). Sequences of dynamic interaction take place through multiple semiotic, prosodic, motor, sensory, and physiological "push and pull." The care partners are usually aware of just a minor portion of a *vast expanse of heterogeneous communication*. These semiotic streams are emotional or procedural communication, implicit memory, and knowledge (Fonagy, 1999, 2002). Much of this semiotic flow is a-conscious, i.e., different from the classical Freudian representational unconscious (Kandel, 2004).

Daniel Stern has emphasized how most therapeutic change occurs in the domain of implicit and procedural knowledge. In his view, the therapeutic process evolves through nonlinear and often unpredictable changes. Therapeutic landmarks can emerge from the experience of *present moments* in which participants create fields of implicit understanding and sharing in the "here and now." In these intense moments, the experience of time duration extends, with a sense of *integrity* in what is happening. The flow of these events will repeat in an evolving spiral creating wider *opportunities for new self-organization* and consolidating changes in implicit memory (Stern, 2004). Explicit verbalizations of these states are not necessary and sometimes even counterproductive. The core of a change process is synchronization and coevolution, generating *dyadic states of expanded and shared consciousness* to be followed by state transitions to new decoupling. From critical states full of expectations and instabilities, waves of change can emerge as new self-organizations (Bak & Sneppen, 1993).

3 Results. Deterministic and Probabilistic Models

A vast range of hidden regulators potentially involved in human synchronizations generates implicit knowledge. For example, subliminal messages make *every conversation an implicit musical* of singing and dance (Shockley et al., 2003). A relevant factor is the intrinsic musicality of spoken language, also called *prosody*. Voice tonality, accent, volume, cadence, flow, pressure, hue, sentence structure, grammar, vocabulary, conversation shifts, and breaks, all these linguistic variables express embodiment in the psychotherapeutic setting (Ryan, 2005). Many studies show how these signals, called *hidden regulators,* can lead to syncing physiological functions, biological rhythms, and hormonal cycles (Hofer, 1981, 1994). We should also consider the role of the *olfactory system* with its direct impact on the entorhinal cortex (Freeman, 1998; Orsucci, 2000). The neural system of mirroring and understanding, called *mirror neurons,* will concur with these hidden regulators (Rizzolatti & Arbib, 1998). Therefore, the deep embodied human coupling will modify functional and organic pathologies (Orsucci, 1996).

Our research laboratory has contributed to studying prosodic structures and how prosodic resonances or dissonances can lead to coupling or decoupling during the therapeutic process. We also developed complementary work on synchronizing physiological variables that indicate emotional responses between partners in psychotherapy (Orsucci, 2021). The *clusters analysis* and *Markov chain* we explored showed how synchronization could happen partially or globally, symmetric and asymmetric.

Resonance and synchronization can generate dynamical landscapes of mixed states, where we might find areas of synchronization, areas of resonance, and areas incoherently drifting away from these coupling phenomena (Kuramoto & Battogtokh, 2002; Abrams & Strogatz, 2004). It would be helpful to differentiate the diachronic mixed states denominated by Kuramoto as the *Chimera States.* In these states, synchronization is diachronic, developing from the entrainment of local frequencies and phases (also called Arnold's tongues) toward larger areas and eventually whole systems formed by networks of oscillators (with their subsystems). These processes are interrelated with the synchronic hybrid dynamical mapping of subsystems organized in different dynamical structures, as highlighted in Schreiber's plot (Schreiber, 1999).

These intricate semiotic dynamics correlate to brain dynamics. Cognitive tasks constantly require an intricate balance between segregated and integrated neural processing (Shine et al., 2016). The relative level of functional integration versus segregation of cognitive systems significantly affects cognitive performance. Highly segregated systems enable efficient computations in local, functionally specialized brain regions. Integrated systems provide rapid consolidation of information across systems, necessary for the coordinated cohesive performance of complex tasks. These crucial brain states are captured in the present framework simultaneously as metastable (segregated) and coherent (integrated) states and, perhaps, the most critical for brain function, like the chimera state that describes partial synchrony across subsets of cognitive systems. Previous work has found that functionally segregated states involve shorter local connections (Liégeois et al., 2016). Integration, instead, largely relies on the global influence of subcortical regions and cortical hubs with many diverse connections to other brain regions (Shine et al., 2018). Recognizing chimera dynamics in phase transitions and dynamic maps can help clarify the hybrid complexity of synchronization in critical cognitive states where a balance between integration and segregation works for adaptive cognition and social interactions (Chouzouris et al., 2018).

Many biological and neural systems are networks of interacting systematic processes. Their functionality depends on the emerging collective dynamics of networks. Recently introduced approaches, known as the Ott–Antonsen and Watanabe–Strogatz reductions, simplify the analysis by bridging small and large scales. Reduced model equations describe the network dynamics with few collective variables. The resulting equations are next-generation models; rather than being heuristic, they can exactly link microscopic and macroscopic descriptions. At the same time, they are sufficiently simple without great computational effort (Bick & Martens, 2015; Bick et al., 2020).

We could define different levels of resolution capturing microscopic, mesoscopic, or macroscopic dynamics of modeling. We will focus on mesoscopic dynamics and macroscopic dynamics and highlight their mapping. We will also consider deterministic versus probabilistic approaches. Finally, we might consider cluster variables influencing the stability of relational dynamics. At the mesoscopic level, this data analysis and mapping highlight dynamical landscapes of mixed coupling, with zones of synchronization, non-interaction, and uncoupling drifts. Moreover, the dynamical mapping can change over time (Orsucci & Tschacher, 2022, 2023).

The Japanese physicist Kuramoto and Kuramoto (1984) proposed a paradigmatic mathematical model to describe synchronization dynamics in a large set of coupled oscillators. The most frequent form of the model has the following equation:

$$\frac{\mathrm{d}\theta_i}{\mathrm{d}t} = \omega_i + \frac{K}{N}\sum_{j=1}^{N}\sin\left(\theta_j - \theta_i\right), \quad i = 1\ldots N,$$

The system consists of N limit-cycle oscillators with phase θ_i and coupling K.

Then, in November 2002, Yoshiki Kuramoto and Dorjsuren Battogtokh published the paper "Coexistence of Coherence and Incoherence in Nonlocally Coupled Phase Oscillators" (Kuramoto & Battogtokh, 2002; Smirnov et al., 2017). They found coexistence of coherence and incoherence in a network of identical nonlocally coupled complex Ginzburg–Landau oscillators. Coupled non-identical oscillators exhibited mixed complex behavior (frequency locking, phase synchronization, partial synchronization, and incoherence). Identical oscillators, instead, were expected to either synchronize in phase or incoherently drift. They showed that oscillators coupled with identical natural frequencies could behave differently from one another for specific initial conditions. Some could synchronize, while others remained incoherent in a stable state. They considered the following equation, which they called the nonlocally coupled complex Ginzburg–Landau equation:

$$\frac{\delta}{\delta t}\psi(x,t) = \omega(x) - \int G\left(x - x'\right)\sin\left(\psi(x.t) - \psi\left(x',t\right) + \alpha\right)\mathrm{d}x'$$

With $\omega(x) = \omega$ for all x.

Later, Abrams and Strogatz (2004); Panaggio & Abrams, (2015) named it a chimera state from the name of the mythological Greek creature made up of parts of different animals and introduced some theoretical clarifications for such behavior. Finally, they studied the most straightforward system presenting a chimera state, a ring of phase oscillators governed by:

$$\frac{\vartheta \phi}{\theta t}\omega - \int_{-\pi}^{\pi} G(x - x') \sin[\phi(x, xt) - \phi(x', t) + \alpha]\mathrm{d}x'$$

Here is the phase of the oscillator at time t. The space variable runs from $-\pi$ to π with periodic boundary conditions. The frequency ω plays no role in the dynamics; one can set $\omega = 0$ by redefining it without changing the equation's form.

Chimera states were later found in limit-cycle oscillators, chaotic oscillators, chaotic maps, and neuronal systems. In the beginning, chimera patterns were observed in nonlocally coupled networks, but these states were also found in globally and locally (nearest neighbor) coupled networks and in modular networks (Schöll et al., 2019; Wang & Liu, 2020). The usage of Markov chains for mapping couplings and chimera states was also explored (Cavers & Vasudevan, 2015; Vasudevan et al., 2015). C. R. Laing studied chimera states in heterogeneous networks analyzing the influence of heterogeneous coupling strengths. Of further interest to human dynamics is the emergence of chimera states in multiscale networks resulting from networking different networks (Laing, 2009; Makarov et al., 2019). Chimera synchronization mapping also applies to coupling non-identical oscillators in hybrid and multiscale networks.

Studies in the dynamic mapping of heterogeneous synchronization indicate that similar dynamics involving different brain areas related to emotional, motor, and verbal interactions co-occur. Cognitive tasks constantly require a balance between segregated and integrated neural processing (Shine et al., 2016) with relevant consequences for cognitive performance. Segregation enables efficient computations in specialized brain regions, while integrated systems ensure coordinated robust performance. Integrated states tend to involve shorter local connections (Liégeois et al., 2016). Integration largely relies on subcortical regions and cortical hubs with diverse connections to other brain regions (Shine et al., 2018). "Recognizing chimera dynamics can help to clarify the hybrid complexity of synchronization in critical cognitive states where a balance between integration and segregation is required for adaptive cognition and social interactions" (Chouzouris et al., 2018). Brain chimera dynamics might also be related to various neuronal interactions mediated by different electrical or chemical synapses in the nervous system. Neural interactions are hybrid also because they involve neuromodulators and hormones, with faster or slower action and different time frames (Majhi et al., 2019), further facilitating the emergence of chimera states (Makarov et al., 2019). Therefore, as different regions dynamically interact to perform neurocognitive tasks, variable patterns of partial synchrony form chimera states (Bansal et al., 2019) (Fig. 1).

We would like to understand how the interplay of network properties (for example, coupling topology, frequency, and strength of interactions) and the characteristics of

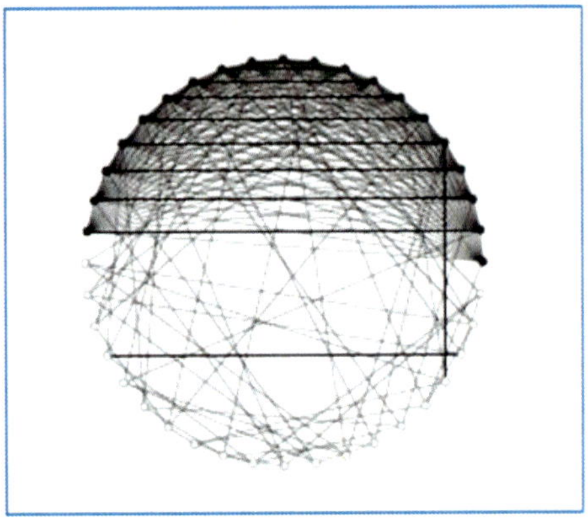

Fig. 1 An ideal example of heterogeneous network: two groups of densely and sparsely connected nodes (Li & Saad, 2017)

the individual nodes shape the emergent dynamics. First, we must consider that coupling and synchrony can come in wide varieties, including phase synchrony, where the state of different oscillators aligns precisely, or frequency synchrony, where the oscillators' frequencies coincide. Some oscillators can respond quickly and change, while others might be slower. Some might have multiple modulators, and others might be more independent. Moreover, synchrony may be global across the entire network or localized in a particular region, the rest of the network being non-synchronized, thus giving rise to synchrony patterns. Therefore, as the textures of synchrony can be highly intricate, a common approach is to streamline the description of each oscillatory node to its simplest form, a phase oscillator. In the reduced system, we can define the state of each oscillator by a single periodic phase variable that captures the state of the periodic process (Bick et al., 2020). Some recent approaches clarify how it is possible to replace a large number of nodes with a collective or mean-field variable that describes the network evolution, reducing the complexity of the problem. Mean-field approaches have a long history of statistical physics in approximating the dynamics of a large ensemble of units. For example, the Ott–Antonsen reduction (Ott & Antonsen, 2008, 2009) and the Watanabe–Strogatz (1993, 1994) reduction can be employed for large or even infinite networks of oscillators.

Macroscopic modeling has addressed interpersonal dynamics in social relations. Interpersonal relationships appear in many contexts, such as family, kinship, acquaintance, work, social media, and epidemiology. The manifestation of interpersonal relationships in society comes in many forms ranging from romantic, parent–child, friendships, comradeship, casual, friend-with-benefits, soulmates, and dating websites. Research mainly focused on three main lines: collaboration/competition, romantic/love, network dynamics in social media, and epidemiology. We will focus on some of the most exciting developments in love and collaboration dyadic interactions.

One of the most intriguing interpersonal relationships in human social life is the romantic relationship (Barley & Cherif, 2011). The mathematical models capturing the dynamics of love between two people recently gained attention among researchers. In a one-page influential work (Strogatz, 1988) and later in a book (Strogatz, 1994), Strogatz applied linear differential equations to study love affairs. Rinaldi (1998; Rinaldi et al., 2015), Sprott (2004, 2008), Liao and Ran (2007), and Wauer et al. (2007) have investigated natural perturbations and extensions of the Strogatzian model by including features such as attraction factors, delay, and nonlinear return functions, or love triangles. Models of this kind are conceptual because they explain a given behavioral property of an entire class of systems. These models are also called minimal or toy models to show that they involve only a few variables and do not claim to explain details too technically (Rinaldi et al., 2015). The most frequently used conceptual models are composed of a finite number of Ordinary Differential Equations (ODEs), one for each variable.

Using the typology of Strogatz (1988, 1994), Sprott (2008), the four romantic styles are summarized below:

(I) Eager Beaver: individual one is encouraged by his feelings (α) as well as that of individual 2 (β) ($\alpha_i > 0$ and $\beta_i > 0$).
(II) Secure or Cautious lover: individual 1 retreats from his feelings but is encouraged by individual 2 ($\alpha_i < 0$ and $\beta_i > 0$).
(III) Hermit: individual 1 retreats from his feelings and that of individual 2 ($\alpha_i < 0$ and $\beta_i < 0$).
(IV) Narcissistic Nerd: individual 1 wants more of what he feels but retreats from the feelings of individual 2 ($\alpha_i > 0$ and $\beta_i < 0$) (Fig. 2).

This classification allows us to characterize the dynamics of various combinations of romantic styles. Previously mentioned papers have considered various dynamics using all possible combinations in parameters α_i and β_i. In this paper, we focus primarily on dynamics observed when individuals in the region (II) interact with individuals in the region (III). We later focus on the stochastic model's features that are not present in the deterministic dynamics. For instance, Cherif et al. (2009) numerically showed that stochastic models of romantic relationships could exhibit exotic

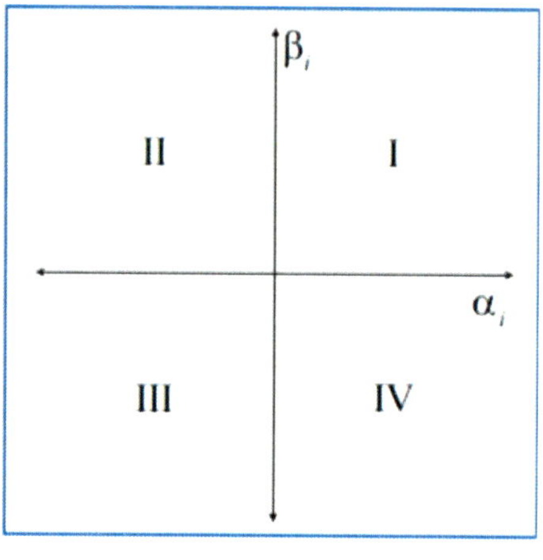

Fig. 2 Romantic styles graph (based on Strogatz, Rinaldi, and Sprott)

dynamics such as sustained oscillations (e.g., stochastic resonance) and diffusion of trajectories (e.g., multiple crossing of stability regions containing stable equilibria), while deterministic models do not and may only show damped oscillations. Therefore, this section presents deterministic (linear and nonlinear) models followed by their stochastic equivalents:

$$\frac{dX_1}{dt} = -\propto_1 X_1 + \beta_1 X_2\left(1 - \varepsilon X_2^2\right) + A_1$$
$$\frac{dX_2}{dt} = -\propto_2 X_2 + \beta_2 X_1\left(1 - \varepsilon X_1^2\right) + A_2$$

The equations above assume that feelings decay exponentially if not activated by contact with partners. For $(X_{1,}X_2) \in R \times R$, where $\alpha_i > 0$ is non-negative, and β_i and A_i $i = 1, 2$ are real constants, respectively. These parameters are *oblivion, reaction,* and *attraction* constants, respectively. For β_i and A_i, we relax the positivity condition. The parameters specify the romantic style of individuals 1 and 2. For instance, α describes the extent to which individual 1 is encouraged by his/her feeling. In other words, it indicates the degree to which an individual has a sense of confidence or anxiety about others' validation. The parameter β represents the extent to which an individual is encouraged by his/her partner and expects his/her partner to be supportive, seeking or avoiding closeness in a romantic relationship. Barley and Cherif (2011) proposed these nonlinear dynamics, while for $\varepsilon = 0$, the model reduces to those proposed by Strogatz and others.

In the previous mathematical papers, all relationship factors are independent. They consider time-invariant personalities and the appeal of individuals, ignoring aging, learning, adaptation, fluctuation of feelings, external interferences from families, other possible attractions, and socio-economic factors. In addition, the role of neuroscience, cultural and institutional conditions, and attachment dynamics are largely ignored. We can consider stochastic variations in previous models by assuming all these forces as external factors. Barley and Cherif (2011) investigated the probabilistic nature of romantic dynamics considering stochastic factors, including additive random drift and periodic parametric excitation (Wauer et al., 2007). In the following section, we provide a method of using a deterministic formulation to derive a stochastic model. The stochastic model is based on a probability continuous-time Markov process, taking the expected changes and a covariance matrix for the Markov process.

(i) Listing all the possible changes $\Delta X = [\Delta X_1, \Delta X_2]$ along with the probabilities for each change in a short time step Δt;
(ii) Taking the expected changes $E\,[\Delta X]$ and covariance matrix $E\,[\Delta X\,(\Delta X)^T]$ are calculated for the Markov process.

Alternatively, we can also arrive at the stochastic model by dividing the expected changes and the square root of the covariance matrix by Δt. In the limit as $\Delta t \to$

0, the former becomes the drift term μ (t, X_1, X_2), and the latter becomes the diffusion coefficient D (t, X_1, X_2). Both procedures yield similar stochastic differential equations of the form:

$$\mathrm{d}X = \mu(t, X_1, X_2)\mathrm{d}t + D(t, X_1, X_2)\mathrm{d}W$$

$W = [W_1, \ldots, W_4]^T$ is an independent Wiener process (Brownian motion). Weak approximations could justify moving from a discrete to a continuous stochastic process. This weak approximation is equivalent to converging a family of discrete state-space Markov chains to a continuous stochastic process.

The stochastic differential equations describing the dynamics of interpersonal and romantic relationships are as follows:

$$\begin{aligned}\mathrm{d}X_1 = {} & \left[-\alpha_1 X_1 + \beta_1 X_2\left(1 - \varepsilon X_2^2\right) + A_1\right]\mathrm{d}t \\ & - \sqrt{\alpha_1} X_1 \mathrm{d}W_1 + \sqrt{\beta_1} X_2\left(1 - \varepsilon X_2^2\right) + A_1 \mathrm{d}W_2\end{aligned}$$

$$\begin{aligned}\mathrm{d}X_2 = {} & \left[-\alpha_2 X_2 + \beta_2 X_1\left(1 - \varepsilon X_1^2\right) + A_2\right]\mathrm{d}t \\ & + \sqrt{\beta_2} X_1\left(1 - \varepsilon X_1^2\right) + A_2 \mathrm{d}W_3 - \sqrt{\alpha_2} X_2 \mathrm{d}W_4\end{aligned}$$

where W represents independent standard Wiener processes, this approach allows us to extend deterministic models to stochastic models. One can analyze the dynamics of the stochastic models with the help of the stability analysis of the deterministic equations. The solution to the deterministic model corresponds to the mean of the stochastic model. The stochastic system exhibits sustained oscillations, whereas a deterministic model shows damped oscillations.

Nowak has extensively studied the role of human cooperation in stabilizing dyadic and network dynamics (Nowak, 2006, 2013). He noticed that direct reciprocity could only lead to the evolution of cooperation if the probability w of another encounter between the same two individuals exceeds the cost-to-benefit ratio of the altruistic act: $w > c/b$.

However, random contextual factors can easily disrupt direct reciprocity. Therefore, indirect reciprocity can introduce additional stabilizing social reputation and reward factors. Thus, cooperation allows for playing a more valuable game, while defection leads to less valuable evolutionary games, as analyzed in stochastic games (Hilbe et al., 2018).

Attachment style can affect cooperation, while establishing a positive working alliance can influence relational dynamics. As a form of cooperation within a therapeutic relationship, the concept of the therapeutic alliance appeared in Freud's (1912/1958) discussion of the various types of transferences. Freud defined transference as the template (or templates) that guides an individual's erotic life. Although Sterba (1934, 1940), Fenichel (1941), Zetzel (1956), Stone (1961), and Gitelson (1962) all took up these ideas in various forms, the term working alliance emerged when Greenson first used it in 1967 (Horvath & Symonds, 1991). Greenson (1967), who described the working alliance as a "rational relationship between patient and analyst" (p. 46), argued that this positive collaboration between therapist and patient is crucial for effective treatment. Subsequent writers have alternated between the terms working alliance and therapeutic alliance. In his comprehensive concept of the working alliance, Bordin (1979) outlined three significant components: (a) agreement on goals for treatment, (b) agreement on tasks to achieve those goals, and (c) the emotional bond of trust and attachment that develops between therapist and patient. A thorough meta-analysis review of this issue proved that individuals with secure attachment styles tend to establish stronger alliances, while insecure attachment would negatively correlate with cooperation (Diener & Monroe, 2011; Bernecker et al., 2014; Marmarosh et al., 2014).

4 Discussion and Conclusions

A deterministic model exhibited damped oscillations with specific parameter values. Stochastic models showed sustained oscillations, the diffusivity of equilibria, and multiple stability boundary crossings with the same parameter values. The results show that deterministic linear and nonlinear models approach locally stable emotional behaviors. However, in the presence of stochasticity in the models, we could observe complex and exotic patterns of emotional behaviors. The stochastic differential equation extension provides insight into the dynamics of human relationships that deterministic models do not capture. Deterministic models assume that individuals respond predictably to their feelings and that of others without external influences, such as ecological factors. Stochastic models capture the fluctuations due to relational dynamics and external influences. The behavioral patterns observed in a stochastic model represent how stable relationship periods are subject to emotional, psychological, social, and cultural effects.

This paper provides new directions to the study of interpersonal relationships. The most fruitful direction from a mathematical perspective is developing methods to analyze systems that exhibit multiple "stability boundary-crossing" or "jump between locally stable equilibria" dynamics. The analysis of these dynamics is worth studying in another paper (Orsucci & Tschacher, 2023).

Relationship coevolution can also derive interesting points from game theory (Nowak, 2006). For example, stochastic game theory shows that establishing cooperation allows for playing a more valuable game, while defection leads to a less valuable game in the next round (Hilbe et al., 2018). The analysis of this idea requires the theory of stochastic games and its introduction into evolutionary game theory. Surprisingly, the dependency on the public resource in previous interactions can significantly enhance the propensity for cooperation.

In the past few decades, the issue of emergence in semiotic dynamics has mobilized researchers from many disciplines and undergone spectacular development (Gong et al., 2014). Emergent macroscale phenomena result from microscale interactions producing new structures and features (Elman, 1996; Orsucci, 2002). In evolutionary linguistics, this term has two distinct senses: (i) ontogenetically, it refers to language acquisition; (ii) phylogenetically, it refers to language origin and bifurcations. This view of language emergence developed to answer some problems faced by innatism (McWhinney, 2001). Emergentism proposes that the existence of linguistic universals does not necessarily imply their prefiguration in the human brain. Instead, universals may have emerged spontaneously and independently in evolving language in response to learning constraints or universal biases during selection. In a "mosaic" fashion, the emergence of semantics, phonology, morphology, and syntax may have occurred at different times and following different routes. Emergentism focuses on questions about how language is the way it is and how various factors shape linguistic structures. Language is assumed to be a kind of interface among various fundamental abilities, some of which may underlie non-linguistic processes, show up to a certain extent in other animals, and emerge probably much earlier than language. Language is a complex reconfiguration of conventional systems adapted in evolutionarily novel ways. Instead of arising because of genetic mutations unrelated to other aspects of human cognition and social life, language is shaped by human cognitive mechanisms and pressures from acquisition and use and arises from an elaboration of domain-general abilities into linguistic activities.

Language is a core component in building the human eco-cognitive niche. In studies of the populations of two species, the Lotka–Volterra system of equations has been extensively used to describe the behavior dynamics between two species, N_1 and N_2. Examples include real-life relations and theoretical analysis of the system's behavior (Nowak, 2006, 2013). The same equations apply to semiotic dynamics as the core component of niche ecology. Across many animal taxa, individuals from groups collectively process information and make joint decisions (Freeman & Orsucci, 2017; Sasaki & Biro, 2017). These groups can generate better decisions than solitary agents by pooling information—a phenomenon referred to as *collective intelligence*. Studies of collective intelligence have typically focused on one-off performance, such as a swarm of bees choosing to settle at one of several available nest sites or a group of jurors reaching a verdict in court. Some studies have generally assumed that those collective decisions are not influenced by information acquired from past experiences involving similar scenarios. However, because animal groups in nature often face the same tasks repeatedly, feedback from past outcomes has the potential to influence future behavior and decision quality. Such "collective learning" (also called culture, knowledge, or science) allows individuals to acquire valuable information through collaborating with others and may provide crucial input when the same task is undertaken by the group again. Iterative solving of a given task may thus lead to accumulating knowledge within the group and, in turn, improve collective performance over time. Improvements in behavioral solutions to specific social, historical, or ecological problems are also central to the study of culture. The accumulation of knowledge through individual invention and subsequent social learning has been recognized as an essential advantage of group living. In addition, culture can accumulate progressively, or "ratchet," over generations—a process called Cumulative Cultural Evolution (CCE) (Caldwell & Millen, 2009). CCE can allow groups to develop increasingly complex knowledge and skills over time beyond the capacities of a single individual.

We accumulated empirical evidence and dynamical models of a seminal intuition generated by Merleau-Ponty (1960), as in essence "*human embodiment is an intricated texture of experience and language.*"

Acknowledgements We would like to send our special thanks to the following colleagues who contributed, directly and indirectly to the development of research and models and ideas. However, we are solely responsible for how they have been presented in this publication. In alphabetical order, we wish to thank, Wolfgang Tschacher, Günter Schiepek, Mario Reda, Giulio Defelice, Omar Gelo, Sergio Salvatore, Alessandro Giuliani, Walter J. Freeman, Giorgio Parisi, Joe Zbilut, Chuck Webber, Eliano Pessa, Sergio Rinaldi, Mario Fulcheri.

References

Abrams, D. M., & Strogatz, S. H. (2004). Chimera states for coupled oscillators. *Physical review letters, 93*(17), 174102.

Andreasen, N. C., & Grove, W. M. (1986). Thought, language, and communication in schizophrenia: Diagnosis and prognosis. *Schizophrenia Bulletin, 12*(3), 348.

Bak, P., & Sneppen, K. (1993). Punctuated equilibrium and criticality in a simple model of evolution. *Physical Review Letters, 71*(24), 4083–4086.

Baker, G. (1998). The private language argument. *Language and Communication, 18*(4), 325–356.

Bansal, K., Garcia, et al. (2019). Cognitive chimera states in human brain networks. Science Advances, 5(4), eaau8535.

Barabasi, A., & Frangos, J. (2003). *Linked: The new science of networks.* Basic Books.

Baranger, M., & Baranger, W. (1985). La situation analytique comme champ dynamique. *Revue Française de Psychanalyse.*

Baranger, M., & Baranger, W. (2008). The analytic situation as a dynamic field. *The International Journal of Psychoanalysis, 89*(4), 795–826.

Barley, K., & Cherif, A. (2011). Stochastic nonlinear dynamics of interpersonal and romantic relationships. *Applied Mathematics and Computation, 217*(13), 6273–6281.

Barlow, M. (1986). The Voynich manuscript-by Voynich? *Cryptologia, 10*(4), 210–216.

Bernecker, S. L., Levy, K. N., & Ellison, W. D. (2014). A meta-analysis of the relation between patient adult attachment style and the working alliance. *Psychotherapy Research, 24*(1), 12–24.

Bick, C., & Martens, E. A. (2015). Controlling chimeras. *New Journal of Physics, 17*(3), 033030.

Bick, C., Goodfellow, M., Laing, C. R., & Martens, E. A. (2020). Understanding the dynamics of biological and neural oscillator networks through exact mean-field reductions: A review. *The Journal of Mathematical Neuroscience, 10*(1), 1–43.

Biro, D., Sasaki, T., & Portugal, S. J. (2016). Bringing a time-depth perspective to collective animal behaviour. *Trends in Ecology and Evolution, 31*(7), 550–562.

Blackmore, S. J. (2000). *The meme machine.* Oxford University Press.

Boccaletti, S., Pecora, L., & Pelaez, A. (2001). Unifying framework for synchronization of coupled dynamical systems. *Physical Review E, 63*(6 Pt 2), 66–78.

Bordin, E. S. (1979). The generalizability of the psychoanalytic concept of the working alliance. *Psychotherapy: Theory, research & practice, 16*(3), 252.

Borges, J. L. (1964). *Labyrinths selected stories and other writings.* Augmented ed. New Directions Pub. Corp.

Brumbaugh, R. S. (1975). The solution of the Voynich 'Roger Bacon' Cipher. *Yale Library Gazette, 49*, 347–355.

Brumbaugh, R. S. (1978). *The most mysterious manuscript: The Voynich "Roger Bacon" cypher manuscript.* Southern Illinois Univ Pr.

Caldwell, C. A., & Millen, A. E. (2009). Social learning mechanisms and cumulative cultural evolution: Is imitation necessary? *Psychological Science, 20*(12), 1478–1483.
Carroll, T. L., & Pecora, L. M. (1993). Stochastic resonance and crises. *Physical Review Letters, 70*(5), 576–579.
Cherif, A., Barley, K., Xiu-Lian, X., Chun-Hua, F., Ai-Fen, L., Da-Ren, H., Helbing, D., Johansson, A., Szabo, G., & Szolnoki, A. (2009). Stochastic nonlinear dynamics of interpersonal and romantic relationships. *arXiv Preprint* arXiv:0911.0013. http://my.arxiv.org/arxiv/FilterServlet/pdf/0911.0013
Chouzouris, T., Omelchenko, I., Zakharova, A., Hlinka, J., Jiruska, P., & Schöll, E. (2018). Chimaera states in brain networks: Empirical neural vs modular fractal connectivity. *Chaos: An Interdisciplinary Journal of Nonlinear Science, 28*(4), 045112.
Clemens, R. (ed). (2016). The Voynich Manuscript (New Haven, Conn./London: Yale University Press Tschacher, W., & Bergomi, C. (2015). The implications of embodiment: Cognition and communication. Andrews UK Limited.
Condray, R., Steinhauer, S. R., van Kammen, D. P., & Kasparek, A. (2002). The language system in schizophrenia: Effects of capacity and linguistic structure. *Schizophrenia Bulletin, 28*(3), 475.
Crow, T. J. (2000). Schizophrenia as the price that Homo sapiens pays for language: A resolution of the central paradox in the origin of the species. *Brain Research Reviews, 31*(2–3), 118–129.
D'Imperio, M. E. (1978). *The Voynich manuscript: An elegant enigma.* Aegean Park.
Dawkins, R. (2006). *The selfish gene.* Oxford University Press.
De Berardis, D., Serroni, N., Campanella, D., Carano, A., Gambi, F., Valchera, A., Conti, M. C.
DeLisi, L. E. (2001). Speech disorder in schizophrenia: Review of the literature and exploration of its relation to the uniquely human capacity for language. *Schizophrenia Bulletin, 27*(3), 481.
Diener, M. J., & Monroe, J. M. (2011). The relationship between adult attachment style and therapeutic alliance in individual psychotherapy: A meta-analytic review. *Psychotherapy, 48*(3), 237.
Dittrich, W. H., Troscianko, T., Lea, S. E., & Morgan, D. (1996). Perception of emotion from dynamic point-light displays represented in dance. *Perception, 25*(6), 727–738.
Dube', J., & Despres, P. (2000). The control of dynamical systems—Recovering order from chaos. In *The physics of electronic and atomic collisions.* AIP.
Durham, W. H. (1992). *Coevolution: Genes, culture, and human diversity.* Stanford Univ Pr.
Eliot, T. S. (2011). *The complete poems and plays of TS Eliot.* Faber & Faber.
Elman, J. L. (1996). *Rethinking innateness, a connectionist perspective on development.* MIT Press.
Freud, S. (1912). The dynamics of transference. *Classics in psychoanalytic techniques, 12,* 97–108.
Fonagy, P. (1999). Memory and therapeutic action. *International Journal of Psycho-Analysis, 80*(2), 215–223.
Fonagy, P. (2002). *Affect regulation, mentalization, and the development of the self.* Other Press.
Fonagy, P., Gergely, G., & Jurist, E. L. (2004). *Affect regulation, mentalization, and the development of the self.* Karnac Books.
Freeman, W. (1975). *Mass action in the nervous system, examination of the neurophysiological basis of adaptive behavior through the EEG.* Academic Press.
Freeman, W. J. (1995). *Societies of brains, study in the neuroscience of love and hate. International neural networks society series.* Lawrence Erlbaum Associates.
Freeman, W. J. (1998). The neurobiology of multimodal sensory integration. *Integrative Physiological and Behavioral Science, 33*(2), 124–129.
Freeman, W. J. (1999). *How brains make up their minds. Maps of the mind.* Weidenfeld & Nicolson.
Freeman, W. J., & Orsucci, F. F. (2017). Semiotic dynamics. Conversations and reflections at Trinity College. *Intentional Neurodynamics in Transition: The Dynamical Legacy of Walter Jackson Freeman, Special Issue of Chaos and Complexity Letters, 11*(1), 183–192.
Fusaroli, R., & Tylén, K. (2016). Investigating conversational dynamics: Interactive alignment, interpersonal synergy, and collective task performance. *Cognitive Science, 40*(1), 145–171.
Fusaroli, R., Rączaszek-Leonardi, J., & Tylén, K. (2014). Dialog as interpersonal synergy. *New Ideas in Psychology, 32,* 147–157.

Gallagher, S. (2001). The practice of mind. Theory, simulation or primary interaction? *Journal of Consciousness Studies, 8*(5–6), 83–108.
Gallagher, S. (2013). The socially extended mind. *Cognitive Systems Research, 25*, 4–12.
George, A. (1990). Whose language is it anyway? Some notes on idiolects. *The Philosophical Quarterly, 40*(160), 275–298.
Gibson, J. J. (1979). *The ecological approach to visual perception*. Houghton Mifflin.
Gong, T., Shuai, L., & Comrie, B. (2014). Evolutionary linguistics: Theory of language in an interdisciplinary space. *Language Sciences, 41*, 243–253.
Gould, S. J., & Eldredge, N. (1977). Punctuated equilibria: The time and mode of evolution reconsidered. *Paleobiology, 3*(2), 115–151.
Greenson, R. R. (1972). Beyond transference and interpretation. Classics in psychoanalytic technique, 97–101.
Guidano, V. (1987). *Complexity of the self*. Guilford Press.
Guidano, V. F. (1991). *The self in process: Toward a post-rationalist cognitive therapy*. Guilford Press.
Hamilton, W. D. (1964). The genetical evolution of social behaviour. *II. Journal of theoretical biology, 7*(1), 17–52.
Heagy, J. F., Carroll, T. L., & Pecora, L. M. (1995). Desynchronization by periodic orbits. *Physical Review E Statistical Physics, Plasmas, Fluids, and Related Interdisciplinary Topics, 52*(2), R1253–R1256.
Higginbotham, J. (2006). Languages and idiolects: Their language and ours. In *The Oxford handbook of philosophy of language* (pp. 140–150).
Hilbe, C., Šimsa, Š, Chatterjee, K., & Nowak, M. A. (2018). Evolution of cooperation in stochastic games. *Nature, 559*(7713), 246–249.
Hofer, M. A. (1981). *The roots of human behavior an introduction to the psychobiology of early development*. Freeman.
Hofer, M. A. (1994). Hidden regulators in attachment, separation, and loss. *Monographs of the Society for Research in Child Development, 59*(2–3), 192–207.
Holstein, J. A., & Gubrium, J. F. (2012). The self we live by: Narrative identity in a postmodern world. *Contemporary Sociology, 31*.
Horvath, A. O., & Symonds, B. D. (1991). Relation between working alliance and outcome in psychotherapy: A meta-analysis. *Journal of counseling psychology, 38*(2), 139.
Kahr, B. (2005). The fifteen critical ingredients of good psychotherapy. In *How does psychotherapy work* (pp. 1–14).
Kandel, E. R. (2004). The molecular biology of memory storage: A dialog between genes and synapses. *Bioscience Reports, 24*(4–5), 475–522.
Kantz, H., & Schreiber, T. (1997). *Nonlinear time series analysis*. Cambridge University Press.
Kraljic, T., Brennan, S. E., & Samuel, A. G. (2008). Accommodating variation: Dialects, idiolects, and speech processing. *Cognition, 107*(1), 54–81.
Kripke, S. A. (1982). *Wittgenstein on rules and private language an elementary exposition*. Harvard University Press.
Kuramoto, Y., & Battogtokh, D. (2002). Coexistence of coherence and incoherence in nonlocally coupled phase oscillators. arXiv preprint cond-mat/0210694.
Kuramoto, Y., & Kuramoto, Y. (1984). *Chemical turbulence* (pp. 111–140). Springer.
Lacan, J. (1978). *The four fundamental concepts of psycho-analysis* (Vol. 1st American ed.). Norton.
Laing, C. R. (2009). Chimera states in heterogeneous networks. *Chaos: An Interdisciplinary Journal of Nonlinear Science, 19*(1), 013113.
Landini, G. (2001). Evidence of linguistic structure in the Voynich manuscript using spectral analysis. *Cryptologia, 25*(4), 275–295.
Langs, R. (1976). *The bipersonal field*. J. Aronson.
Laplanche, J., & Pontalis, J. B. (1974). *The language of psycho-analysis*. Norton.
Li, B., & Saad, D. (2017). Chimera-like states in structured heterogeneous networks. *Chaos: An Interdisciplinary Journal of Nonlinear Science, 27*(4), 043109.

Liao, X., & Ran, J. (2007). Hopf bifurcation in love dynamical models with nonlinear couples and time delays. *Chaos, Solitons and Fractals, 31*(4), 853–865.
Liégeois, R., et al. (2016). Cerebral functional connectivity periodically (de) synchronizes with anatomical constraints. *Brain structure and function, 221*, 2985–2997.
Lorenzer, A. (1977). *Sprachspiel und Interaktionsformen: Vorträge und Aufsätze zu Psychoanalyse, Sprache und Praxis*. Suhrkamp Frankfurt a. M.
Lumsden, C. J., Wilson, E. O., & Osborne, E. (1981). *Genes, mind, and culture: The coevolutionary process*. Harvard University Press.
Mainzer, K. (1994). *Thinking in complexity the complex dynamics of matter, mind, and mankind*. Springer.
Majhi, S., et al. (2019). Solitary states in multiplex networks owing to competing interactions. Chaos: An Interdisciplinary Journal of Nonlinear Science, 29(1).
Makarov, V. V., et al. (2019). Multiscale interaction promotes chimera states in complex networks. *Communications in Nonlinear Science and Numerical Simulation, 71*, 118–129.
Margulis, L., Asikainen, C. A., & Krumbein, W. E. (2011). *Chimeras and consciousness: Evolution of the sensory self*. MIT Press.
Marmarosh, C. L., Kivlighan, D. M., Jr., Bieri, K., LaFauci Schutt, J. M., Barone, C., & Choi, J. (2014). The insecure psychotherapy base: Using client and therapist attachment styles to understand the early alliance. *Psychotherapy, 51*(3), 404.
Maturana, H. R., & Varela, F. J. (1980). *Autopoiesis and cognition the realization of the living*. D. Reidel Pub. Co.
Maturana, H. R., & Varela, F. J. (1992). *The tree of knowledge about the biological roots of human understanding*. Shambhala.
Merleau-Ponty, M. (1960). *Signes*. Gallimard.
McWhinney, B., & MacWhinney, B. (1999). The emergence of language from embodiment. The emergence of language, B. MacWhinney, London: Erlbaum, 213–256.
Newman, M. (2006). *The structure and dynamics of networks*. Oxford University Press.
Nowak, M. A. (2006). *Evolutionary dynamics*. Harvard University Press.
Nowak, M. A. (2013). *Evolution, games, and god*. Harvard University Press.
Orsucci, F. (1981a). Linguistic sign in neurosis and psychosis. *Neuropsychiatric Work, 68*(1–2), 66–78.
Orsucci, F. (1981b). Structural analysis of schizophrenic language. *Neuropsychiatric Work, 68*(1–2), 57–65.
Orsucci, F. (1996). Behçet's disease and psychosomatic patterns of thinking. *Psychotherapy and Psychosomatics, 65*(2), 112–114.
Orsucci, F., et al. (2013). Prosody and synchronization in cognitive neuroscience. *EPJ Nonlinear Biomedical Physics, 1*(1), 1–11.
Orsucci, F. (1998). *The complex matters of the mind*. World Scientific.
Orsucci, F. (2000). *Olfaction: New frontiers for cognitive sciences*. World Scientific Publishing.
Orsucci, F. (2002). *Changing mind. Transitions in natural and artificial environments*. World Scientific.
Orsucci, F. (2008). Ethos in everyday action: Notes for a mindscape of bioethics. In *Developments in chaos and complexity research*. Nova Science.
Orsucci, F. (2008). Reflexing interfaces. In *Reflexing interfaces: The complex coevolution of information technology ecosystems*. IGI Pub.
Orsucci, F. (2009). *Mind force on human attractions*. World Scientific Publishing.
Orsucci, F. (2013). Coevolution and linguistic creation. *Journal of Experiential Psychotherapy/ Revista de Psicoterapie Experiential, 16*(3).
Orsucci, F. (2015). Towards a meta-model of human change, from singularity to event horizon. *Chaos and Complexity Letters, 9*(2), 107.
Orsucci, F. (2021). Human synchronization maps—The hybrid consciousness of the embodied mind. *Entropy, 23*(12), 1569.

Orsucci, F., & Tschacher, W. (2022). Complexity science in human change: Research, models, clinical applications. In *Entropy* (Vol. 24(11), p. 1670).

Orsucci, F., & Tschacher, W. (2023). *Complexity science in human change*. MDPI.

Orsucci, F., Giuliani, A., Webber, C., Mazza, M., & Fonagy, P. (2006). Combinatorics and synchronization in natural semiotics. *Physica A: Statistical Mechanics and Its Applications, 361*(2), 665–676.

Orsucci, F., Walter, K., Giuliani, A., Webber, C. Jr, & Zbilut, J. (1999). Orthographic structuring of human speech and texts: Linguistic application of recurrence quantification analysis. *International Journal of Chaos Theory and Applications, 4*, 2–12.

Ott, E., & Antonsen, T. M. (2008). Low dimensional behavior of large systems of globally coupled oscillators. *Chaos: An Interdisciplinary Journal of Nonlinear Science, 18*(3), 037113.

Ott, E., & Antonsen, T. M. (2009). Long time evolution of phase oscillator systems. *Chaos: An Interdisciplinary Journal of Nonlinear Science, 19*(2), 023117.

Panaggio, M. J., Abrams, D. M., Ashwin, P., & Laing, C. R. (2016). Chimera states in networks of phase oscillators: The case of two small populations. *Physical Review E, 93*(1), 012218.

Pikovsky, A., Rosenblum, M., & Kurths, J. (2001). *Synchronization a universal concept in nonlinear sciences* (Vol. 12). Cambridge University Press.

Pyragas, K. (2001). Control of chaos via an unstable delayed feedback controller. *Physical Review Letters, 86*(11), 2265–2268.

Reda, M. A. (1988). *Complex cognitive systems of psychotherapy*. Carocci.

Reda, M. A., & Baskets, L. (2008). Evaluation of psychophysiological dyads patient/therapist during a psychotherapy session: Preliminary results. In *Studies and research of the department of neurological and behavioural sciences*. Siena Ignatius Press Publisher.

Rinaldi, S. (1998). Love dynamics: The case of linear couples. *Applied Mathematics and Computation, 95*(2), 181–192.

Rinaldi, S., et al. (2015). *Modeling love dynamics* (Vol. 89). World Scientific.

Rizzolatti, G., & Arbib, M. A. (1998). Language within our grasp. *Trends in Neurosciences, 21*(5), 188–194.

Ryan, J. (2005). *How does psychotherapy work?* Karnac Books.

Sasaki, T., & Biro, D. (2017). Cumulative culture can emerge from collective intelligence in animal groups. *Nature Communications, 8*, 15049.

Schreiber, T. (1999). Interdisciplinary application of nonlinear time series methods. *Physics Reports, 308*(1999), 1–64.

Searles, H. F. (1979). *Countertransference and related subjects, selected papers*. International Universities Press.

Shine, J. M., et al. (2018). The modulation of neural gain facilitates a transition between functional segregation and integration in the brain. *eLife, 7*, e31130.

Shockley, K., Santana, M. V., & Fowler, C. A. (2003). Mutual interpersonal postural constraints are involved in cooperative conversation. *Journal of Experimental Psychology, 29*(2), 326–332.

Sprott, J. C. (2004). Dynamical models of love. *Nonlinear Dynamics, Psychology, and Life Sciences, 8*(3), 303–314.

Sprott, J. C. (2008). Predator-prey dynamics for rabbits, trees, and romance. In *Unifying themes in complex systems IV* (pp. 231–238). Springer. http://link.springer.com/chapter/10.1007/978-3-540-73849-7_26

Stern, D. N., Brüschweiler-Stern, N., Harrison, A. M., Lyons-Ruth, K., Morgan, A. C., Nahum, J. P., et al. (1998). Involving the process of therapeutic change implicit knowledge: Some implications of developmental observations for adult psychotherapy. *Infant Mental Health Journal, 19*(3), 300–308.

Stern, D. N. (2004). *The present moment in psychotherapy and everyday life*. Norton.

Strogatz, S. H. (1988). Love affairs and differential equations. *Mathematics Magazine, 61*(1), 35.

Strogatz, S. H. (1994). *Nonlinear dynamics and Chaos with applications to physics, biology, chemistry, and engineering*. Addison-Wesley Pub.

Strogatz, S. H. (2003). *Sync: The emerging science of spontaneous order*. Hyperion.

Trevarthen, C. (1998). The concept and foundations of infant intersubjectivity. *Intersubjective Communication and Emotion in Early Ontogeny, 15*, 46.
Trevarthen, C. (2012). Embodied human intersubjectivity: Imaginative agency, to share meaning. *Cognitive Semiotics, 4*(1), 6–56.
Tronick, E. Z., Brüschweiler-Stern, N., Harrison, A. M., Lyons-Ruth, K., Morgan, A. C., Nahum, J. P., Sander, L., et al. (1998). Dyadically expanded states of consciousness and the process of therapeutic change. *Infant Mental Health Journal, 19*(3), 290–299.
Vasudevan, K., et al. (2015). Earthquake sequencing: Chimera states with Kuramoto model dynamics on directed graphs. *Nonlinear Processes in Geophysics, 22*(5), 499–512.
Vasudevan, K., & Cavers, M. S. (2015, April). Chimera States in Earthquake Sequencing: Preliminary Results from Global Seismic Records. In EGU General Assembly Conference Abstracts (p. 15797).
Watanabe, S., & Strogatz, S. H. (1993). Integrability of a globally coupled oscillator array. *Physical Review Letters, 70*(16), 2391.
Watanabe, S., & Strogatz, S. H. (1994). Constants of motion for superconducting Josephson arrays. *Physica D: Nonlinear Phenomena, 74*(3–4), 197–253.
Wauer, J., Schwarzer, D., Cai, G. Q., & Lin, Y. K. (2007). Dynamical models of love with time-varying fluctuations. *Applied Mathematics and Computation, 188*(2), 1535–1548.
West, S. A., Griffin, A. S., & Gardner, A. (2007a). Evolutionary explanations for cooperation. *Current Biology, 17*(16), R661–R672.
West, S. A., Griffin, A. S., & Gardner, A. (2007b). Social semantics: Altruism, cooperation, mutualism, strong reciprocity and group selection. *Journal of Evolutionary Biology, 20*(2), 415–432.
Wittgenstein, L. (1967). *Philosophical Investigations*. Oxford University Press.

Complexity

Quantum Coding of the Self

Paola Zizzi and Massimo Pregnolato

Abstract Self-awareness disturbance is a feature in many mental illnesses, such as schizophrenia and autism spectrum disorder, and may be due to various triggering factors which have been studied at psychiatric, medical, and biological level. At the level of quantum logic, however, it seems possible to formalize the loss of self-awareness in terms of quantum error-correcting codes. Cases of schizophrenia and autism spectrum disorder are presented, the former strictly in the context of quantum error-correcting codes, the latter in relation to the lack of a quantum metalanguage and, consequently, the lack of coherent states.

1 Introduction

In Memory of Eliano Pessa

The concept of the Self has long been studied and debated in philosophy, psychology, neuroscience, and psychiatry and also in sociology (Johnston & Malobou, 2013) but not in logic. We think that in classical logic, the Self can be described in terms of the law of identity, which is one of the three laws of thought. But it must be remembered that the law of identity is placed in the object language separately from both logical rules and structural rules, since it is an axiom. In the classic case, therefore, the identity axiom does not really belong either to the object language, or to the metalanguage that generates and describes it. In this context, the Self is a proposition whose description is a datum, which does not originate in the metalanguage. In other words, in this case, the Self has no semantics. In the classical case, the Self is completely separated from the "Other" because the classical identity

P. Zizzi (✉)
Department of Brain and Behavioural Sciences, Pavia University, Piazza Botta, 11, 27100 Pavia, Italy
e-mail: paola.zizzi@unipv.it

M. Pregnolato
Department of Drug Science, University of Pavia, Via Taramelli 12, 27100 Pavia, Italy
e-mail: massimo.pregnolato@unipv.it

G. Minati and M. Pietronilla Penna (eds.), *Multiple Systems*, Contributions to Management Science, https://doi.org/10.1007/978-3-031-44685-6_13

axiom divides the universe of discourse into two clear-cut parts, with no intersection between them (a dichotomy).

In classical logic, therefore, the Self is something absolute, which cannot be generated, destroyed, or modified.

The quantum identity axiom instead is constructed in the quantum metalanguage (Zizzi, 2010) by means of assertions with an assertion degree, and two metalogical operators. Consequently, the Self is built, and as we shall see, it can be destroyed. It therefore appears that the loss of self-awareness is a quantum phenomenon.

In this work, we will describe the quantum Self also in set theory, as a fuzzy set.

The quantum Self is a compound logical proposition composed of the Self and the non-Self, linked together by a quantum logical connective (Zizzi, 2010) which resembles the quantum superposition, so we can see its physical interpretation as a qubit.

As such, the quantum Self is an unknown quantum state that can be encoded into a logical state, which can undergo changes, due to quantum errors. For this, we suppose that quantum error-correcting codes could be useful for the treatment of some pathologies of the Self, such as schizophrenia.

On the other hand, by a classification of the identity axiom in three different types: classical, semi-classical, and quantum, we argue that coherent states of light (Leisman et al., 2018; Hamblin, 2022) might be useful to cure autism.

The paper is organized as follows.

In Sect. 2, we show the metalogical origin of the quantum Self, and we describe the latter as a fuzzy set.

In Sect. 3, we encode the unknown quantum Self in a 3-qubit logical state and discuss the possible errors, which can be detected and corrected by the 3-qubit quantum error-correcting code.

In Sect. 4, we discuss the disturbances of conscious awareness, or Self-Disorders (SDs) as a disrupted sense of Self and how self-disturbance manifests in the lives of those living with schizophrenia and autism.

Section 5 is devoted to the conclusions.

2 The Quantum Self

As we will show in this section, the quantum Self originates in the quantum metalanguage and is physically interpreted as an unknown quantum state.

2.1 *The Classical Laws of Though and the Self*

The classical laws of thought are:

Law of identity: $A \to A$ (states that an object is the same as itself).

Law of excluded middle: $(A \vee \neg A) = 1$ (A, or non A is true).

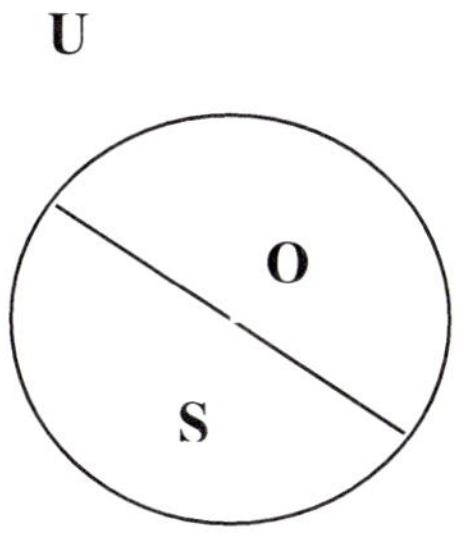

Fig. 1 Classical law of identity. The universe (U) is divided into the self (S) and the other (O)

Law of non-contradiction: $(A \wedge \neg A) = 0$ (A, and non A is false).

The classical law of identity divides the universe of discourse into two parts: the "Self" and the "Other". See Fig. 1.

It is a dichotomy, that is, a partition of the whole into two parts that are:

Mutually exhaustive $S \cup O = U \rightarrow$ (excluded middle).

Mutually exclusive $S \cap O = \emptyset \rightarrow$ (non-contradiction).

The Other is the complement of the Self, $O \equiv S^C$ in U.

2.2 *The Quantum Self from Quantum Metalanguage*

The quantum analog of the law of identity, i.e., the first quantum law of thought is the quantum identity axiom:

$$\boldsymbol{A}\Big|\!-^{|\alpha|^2}\boldsymbol{A} \tag{1}$$

with partial truth value $\boldsymbol{v}_\alpha$ (partial truth values were first introduced in fuzzy logic (Zadeh, 1968) where here $\boldsymbol{v}_\alpha = |\boldsymbol{\alpha}|^2$, and the complex number $\boldsymbol{\alpha}$ is the assertion degree (Zizzi, 2010) of a quantum assertion:

$$\Big|\!-^{\alpha}\boldsymbol{A}\,. \tag{2}$$

The assertion degree α matches to the probability amplitude, and therefore, the partial truth value $\boldsymbol{v}_\alpha$ corresponds to the quantum probability:

$$\boldsymbol{v}_\alpha \equiv \boldsymbol{p} = |\boldsymbol{\alpha}|^2 \in [\mathbf{0}, \mathbf{1}]. \tag{3}$$

From Eqs. (1) and (3), it follows that a quantum object is only probabilistically equal to itself.

The ket $\alpha|\Psi\rangle \equiv |\Psi\rangle_\alpha$ can be associated with the quantum statement (2):

$$\Big|\!-^{\alpha}\boldsymbol{A} \leftrightarrow |\boldsymbol{\Psi}\rangle_{\alpha}. \tag{4}$$

It must be understood that $|\Psi\rangle_{\alpha}$, which depends on the parameter $\boldsymbol{\alpha}$, is not a quantum state, but it is rather a functional on the Hilbert space H with values on the complex numbers, C:

$$\Psi \rightarrow \Psi(\alpha), |\Psi\rangle \in H, \alpha \in C. \tag{5}$$

We recall that a functional is a function defined on a set of functions, F, called functional space; with values in R or in C. Usually on F, we consider structures such as that of vector space or topological space. In our case, the functional space F is a Hilbert space, H, which is a vector space, endowed with an inner product which makes it a fully metric space.

In logical terms, $|\Psi\rangle_{\alpha}$ is the meta-linguistic version of the quantum state $|\boldsymbol{\Psi}\rangle$, the latter being the physical interpretation of the proposition A in the quantum object language:

$$A \equiv^{I} |\Psi\rangle. \tag{6}$$

where "I" stands for "interpretation".

By applying the *-duality (Zizzi, 2010), the quantum version of the Girard–Sambin duality (Sambin et al., 2000; Girard, 1987), to Eq. (2) one gets:

$$(\vert\!-^{\alpha} \boldsymbol{A})^{*} \equiv \boldsymbol{A}\vert\!-^{\alpha^{*}}. \tag{7}$$

In analogy with Eq. (4), the bra $_{\alpha^{*}}\langle\Psi| \equiv \alpha^{*}\langle\Psi|$ is associated with Eq. (7):

$$A\vert\!-^{\alpha^{*}} \leftrightarrow {}_{\alpha^{*}}\langle\Psi|. \tag{8}$$

The use of the "gluing operator" $\circ$ (Zizzi, 2010) on Eqs. (2) and (7) gives back Eq. (1):

$$\boldsymbol{A}\vert\!-^{\alpha^{*}} \circ \vert\!-^{\alpha} \boldsymbol{A} \equiv \boldsymbol{A}\Big\vert\!-^{|\alpha|^{2}} \boldsymbol{A}. \tag{9}$$

Using Eqs. (4) and (8), it follows that the quantum identity axiom in Eq. (1) is associated with the scalar product $_{\alpha^{*}}\langle\Psi||\Psi\rangle_{\alpha}$:

$$A\Big\vert\!-^{|\alpha|^{2}} A \leftrightarrow {}_{\alpha^{*}}\langle\Psi||\Psi\rangle_{\alpha} = |\alpha|^{2}\langle\Psi \mid \Psi\rangle = |\alpha|^{2}. \tag{10}$$

The last equality in Eq. (10) holds because the states are normalized to unity: $\langle\Psi|\Psi\rangle = 1$.

For $|\boldsymbol{\alpha}|^{2} = \mathbf{1}$, we recover the classical identity axiom $\boldsymbol{A}\vert\!-\boldsymbol{A}$.

If the quantum proposition A refers to the Self ($\boldsymbol{A} \equiv \boldsymbol{S}$), then Eq. (2) reads:

"The Self is not asserted completely, but through a certain degree of assertion":

$$\vert\!-^{\alpha} \boldsymbol{S}. \tag{11}$$

By applying the *-duality to the Self in Eq. (11), one gets:

$$S|-^{\alpha^*}. \tag{12}$$

Then, the use of the "gluing operator" ∘ (Zizzi, 2010) on Eqs. (11) and (12): gives:

$$S\Big|-^{|\alpha|^2} S. \tag{13}$$

The conclusion, in Eq. (13), is that the quantum Self is probabilistically the same of itself (the "faded" Self, for example, in schizophrenia (Zizzi & Pregnolato, 2012)).

The quantum superposition of the Self with its primitive negation $S^{\perp}$ is:

$$|-S_{\alpha}\&_{\beta}S^{\perp}. \tag{14}$$

where $\alpha\&_{\beta}$ is a new logical connective (Zizzi, 2010) called "quantum superposition", which is the quantum analog of the logical conjunction "and" ($\wedge$).

The partial truth value v_S of the Self can be given in terms of the partial truth value $v_{S^{\perp}} = |\beta|^2$ of the negated Self, $S^{\perp}$:

$$v_S = |\alpha|^2 = 1 - v_{S^{\perp}} = 1 - |\beta|^2. \tag{15}$$

Notice that the classical case $v_S = 1$ is recovered for $v_S^{\perp} = 0$, that implies that $S^{\perp}$ is the false:

$$S^{\perp}|-\perp. \tag{16}$$

The truth value $v_S = 0$, on the other hand, is not contemplated in the classical case (the law of identity being an axiom is true by definition) but is only a possibility of the quantum case, and would affirm that the Self is the false:

$$S|-\perp. \tag{17}$$

Equation (17) describes the destruction of the Self. We argue then that the loss of self-awareness can only have a quantum description.

Finally, it should be noted that the gluing operator, based on the *duality, is a (quantum) meta-operator, which does not exist in the classical case. This explains why the quantum identity axiom, differently from its classical analog, is a meta-axiom, as it references only to objects of the quantum metalanguage: quantum assertions, and the two meta-operators *-duality and gluing. As we have seen, the quantum Self can be "constructed" (and "destroyed"); therefore, it is a "constructive object", that is, the result of a constructive process. It is neither an "eternal" object nor an absolute truth, and in this sense belongs to the realm of constructivism (Dummett, 1977; Heyting, 1971; Troelstra, 1991).

2.3 The Quantum Self as a Fuzzy Set

So far, we described the quantum Self in logical terms, more precisely, in terms of a multi-valued logic, like, for example, fuzzy logic (Dummett, 1977; Heyting, 1971; Troelstra, 1991). In this sub-section, we will give the set-theoretical counterpart of the logical description given above. In fact, we will describe the quantum Self as a fuzzy set. Fuzzy sets were introduced by Zadeh in 1965 (Zadeh, 1965).

A fuzzy set A is a set whose elements are described by a membership function $f_A(x)$ (where $x \in U$, and U is the universe of discourse), which is valued in the real unit interval [0, 1].

The membership function $f_A(x)$ then associates to every element $x \in U$ a real number in [0, 1]:

$$f_A(x) : \forall x \in U \to [0, 1]$$

A fuzzy set A is formally defined as the set of pairs $[x, f_A(x)]$:

$$A \equiv \{[x, f_A(x)]\}.$$

If $f_A(x) = 0$, x is not included in A.
If $f_A(x) = 1$, x is fully included in A.
If $0 < f_A(x) < 1$, x is partially included in A.

The indicator function $I_A(x)$ of classical sets is a special case of the membership function $f_A(x)$ of fuzzy sets when $f_A(x)$ only takes the values 0 and 1 (crisp set).

By following the above definitions, we will consider the following case of interest.

Let be:

$$x \equiv |\Psi\rangle_\alpha \in U \tag{18}$$

where U is the Hilbert space: $U \equiv H$

The membership function $f_A(x)$ is then a function of $|\Psi\rangle_\alpha$:

$$f_A(|\Psi\rangle_\alpha) : \forall |\Psi\rangle_\alpha \in H \to [0, 1] \tag{19}$$

and the fuzzy set A is defined as:

$$A \equiv \{[|\Psi\rangle_\alpha, f_A(|\Psi\rangle_\alpha)]\} \tag{20}$$

The membership function in Eq. (19) can be defined explicitly as:

$$f_A(|\Psi\rangle_\alpha) : \forall |\Psi\rangle_\alpha \in H \to {}_{\alpha^*}\langle \Psi \mid \Psi\rangle_\alpha = |\alpha|^2 \in [0, 1]. \tag{21}$$

If $|\boldsymbol{\Psi}\rangle_\alpha$ is the meta-interpretation of the quantum Self, i.e., $|-^{\alpha}\boldsymbol{S} \leftrightarrow |\boldsymbol{\Psi}\rangle_\alpha$, then ${}_{\alpha^*}\langle\Psi|$ is the meta-interpretation of the *-dual Self $S|-^{\alpha^*} \leftrightarrow {}_{\alpha^*}\langle\Psi|$, and ${}_{\alpha^*}\langle\Psi \mid \Psi\rangle_\alpha$ is the meta-interpretation of Eq. (13) (the "faded" Self).

The crisp set, with values ${}_{\alpha^*}\langle\Psi \mid \Psi\rangle_\alpha = \{0, 1\}$, is the meta-interpretation of the two limit cases: for value 0, the destroyed Self, and for value 1, the classical Self.

At the light of the above definitions, the quantum Self can be viewed as a fuzzy set S:

$$\boldsymbol{S} \equiv \left\{\left[|\boldsymbol{\Psi}\rangle_\alpha, f_S(|\boldsymbol{\Psi}\rangle_\alpha)\right]\right\} \tag{22}$$

with:

$$f_S(|\boldsymbol{\Psi}\rangle_\alpha) : \forall|\boldsymbol{\Psi}\rangle_\alpha \in \boldsymbol{H} \to [\mathbf{0}, \mathbf{1}]. \tag{23}$$

For a fuzzy set A, its complement $\neg\boldsymbol{A}$, where the symbol $\neg$ represents the negation, (or A^C) is defined by the membership function:

$$\forall \boldsymbol{x} \in \boldsymbol{U} : f_{\neg A}(\boldsymbol{x}) = 1 - f_A(\boldsymbol{x}).$$

In our case, the complement of the Self is the negated self $\boldsymbol{S}^\perp$ in the quantum superposition in Eq. (14). We have then:

$$\forall|\boldsymbol{\Psi}\rangle_\alpha \in \boldsymbol{H} : f_{S^\perp}(|\boldsymbol{\Psi}\rangle_\alpha) = \mathbf{1} - f_S(|\boldsymbol{\Psi}\rangle_\alpha) = \mathbf{1} - |\alpha|^2 = |\boldsymbol{\beta}|^2. \tag{24}$$

The complement c is a standard negator:

$$c(x) = 1 - x$$

which is a strong negator (involutive):

$$cc(x) = x \quad \forall x \in [0.1]$$

with:

$$c(0) \equiv 0^\perp = 1; \quad c(1) \equiv 1^\perp = 0.$$

For the standard negator, the unique fixed point $x*$, such that $c(x*) = x*$, is $x* = 0.5$.

In our case, this is $|\alpha|^2 = 0.5$, corresponding to the Self-cat state, with $\alpha = \beta = \frac{1}{\sqrt{2}}$ in Eq. (14), namely:

$$|-S_{\frac{1}{\sqrt{2}}}\&_{\frac{1}{\sqrt{2}}} S^\perp. \tag{25}$$

Equation (25) means that the Self, *S*, and its negation $S^{\perp}$ have the same weights in the quantum superposition state. The complement of the Self-cat state $S_{\frac{1}{\sqrt{2}}} \&_{\frac{1}{\sqrt{2}}} S^{\perp}$ is:

$$c\left(S_{\frac{1}{\sqrt{2}}} \&_{\frac{1}{\sqrt{2}}} S^{\perp}\right) = \left(S_{\frac{1}{\sqrt{2}}} \&_{\frac{1}{\sqrt{2}}} S^{\perp}\right)^{\perp} = \left(S^{\perp}{}_{\frac{1}{\sqrt{2}}} \&_{\frac{1}{\sqrt{2}}} S\right) = S_{\frac{1}{\sqrt{2}}} \&_{\frac{1}{\sqrt{2}}} S^{\perp} \tag{26}$$

So, Self-cat state is a fixed point. As we will see in the next section, this quantum state is not affected by bit-flip errors, and therefore is the best candidate for mental health.

3 Quantum Error-Correcting Codes and the Self

A universal quantum computer (Benioff, 1980; Deutsch, 1985; Feynman, 1982) could provide enormous computing power, but all attempts to build it have been hampered by noise resulting from runaway interactions between physical qubits and their environment. These quantum errors can be mitigated by quantum error correction (Kitaev, 2003; Steane, 2006; Bombin, 2010; Fowler et al., 2012; Yao et al., 2012; Terhal, 2015).

The latter consists of three phases:

(1) A logical qubit is encoded in the collective state of many physical qubits.
(2) Suitable quantum measurements (syndromes) are used to detect errors.
(3) Appropriate quantum circuits are used to correct the errors detected.

In what follows, we will give a brief introduction to this procedure.

Given the arbitrary qubit $\alpha|0\rangle + \beta|1\rangle$ and *n* physical qubits, the encoded logical qubit $|Q_L\rangle$ will be:

$$|Q_L\rangle = \alpha \underbrace{|0....0\rangle}_{n+1 \text{ times}} + \beta \underbrace{|1....1\rangle}_{n+1 \text{ times}} \tag{27}$$

This can be achieved by initializing the *n* qubits in the ground state $\underbrace{|0\ldots0\rangle}_{n \text{ times}}$.

And write the logical qubit as:

$$|Q_L\rangle = (\alpha|0\rangle + \beta|1\rangle) \otimes \underbrace{|0\ldots0\rangle}_{n \text{ times}} \tag{28}$$

3.1 Three-Qubit Quantum Error-Correcting Code

Let us consider as an example the case $n = 2$. The entanglement process is given by the two CNOT gates as illustrated in Fig. 2.

However, quantum errors can occur, and adequate measurement is required to detect them, without disturbing the quantum superposed state $|\Psi\rangle_L$.

Such measurements, which are called "syndromes", use extra ancilla $|\mathbf{0}\rangle$ (in this case two). The quantum circuit of the syndrome in our case is shown in Fig. 3.

The fourth ancilla is entangled with the first and second qubit, and the fifth ancilla is entangled with the first and third qubit. If there is an error in the block, for example, in the first qubit $|\mathbf{100}\rangle$ there is an odd parity between the first and second qubit, which will turn into a "1" in the control, and the fourth ancilla target will flip $|\mathbf{0}\rangle \rightarrow |\mathbf{1}\rangle$. In $|\mathbf{100}\rangle$, there is also an odd parity between the first and third qubit, which still corresponds to a "1" in the control, and the fifth ancilla target will flip $|\mathbf{0}\rangle \rightarrow |\mathbf{1}\rangle$ as

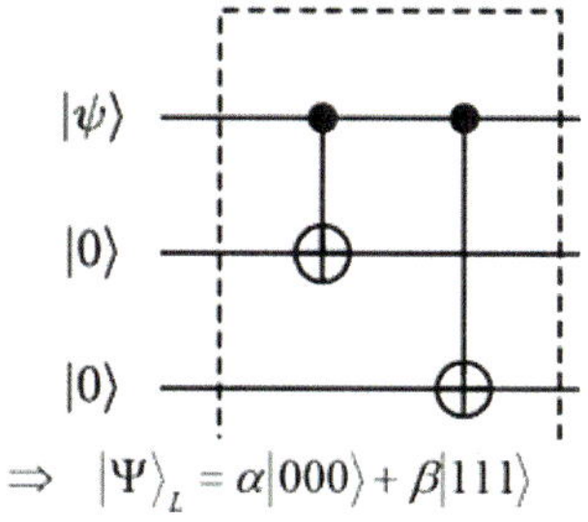

Fig. 2 Quantum circuit for 3-qubit bit-flip code. On the LHS, $|\Psi\rangle$ is the unknown system qubit, which acts as a control, and the two $|0\rangle$ qubits are the targets. On the RHS, the output is the logical qubit $|\Psi\rangle_L$ in case no error has occurred

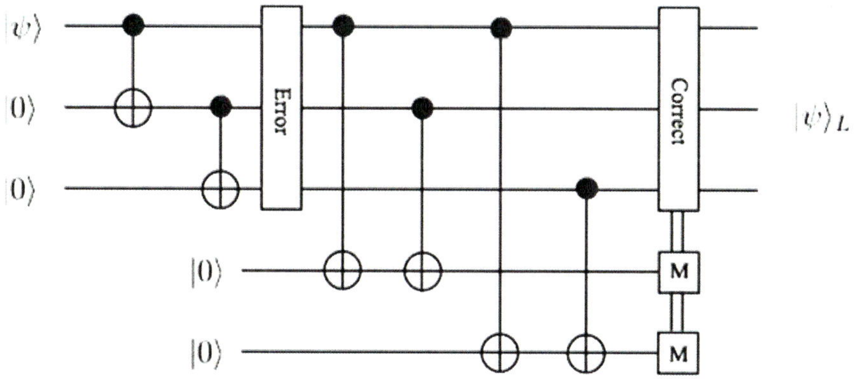

Fig. 3 Quantum circuit of the syndrome with two extra ancilla. The first part of the circuit gives the logical encoded state $|\Psi\rangle_L$; the second part is engineered to detect the possible errors of $|\Psi\rangle_L$

well. Then, the syndrome measurement $|\mathbf{11}\rangle$ signals that there is an error in the first qubit, and a NOT gate should be applied to it to correct the error.

The logical qubit is a vector of the Hilbert space $\boldsymbol{C}^2 \otimes \boldsymbol{C}^2 \otimes \boldsymbol{C}^2$, and the four syndromes form a basis state for the Hilbert space $C^2 \otimes C^2$ of two qubits:

$$
\begin{aligned}
|\boldsymbol{Sy}\rangle_0 &= |\mathbf{00}\rangle \to \boldsymbol{no\ errors}(|\boldsymbol{E}_0\rangle) \\
|\boldsymbol{Sy}\rangle_1 &= |\mathbf{11}\rangle \to \boldsymbol{error} \to_{\leftarrow} \boldsymbol{qubit1}(|\boldsymbol{E}_1\rangle) \\
|\boldsymbol{Sy}\rangle_2 &= |\mathbf{10}\rangle \to \boldsymbol{error} \to_{\leftarrow} \boldsymbol{qubit2}(|\boldsymbol{E}_2\rangle) \\
|\boldsymbol{Sy}\rangle_3 &= |\mathbf{01}\rangle \to \boldsymbol{error} \to_{\leftarrow} \boldsymbol{qubit3}(|\boldsymbol{E}_3\rangle)
\end{aligned} \tag{29}
$$

where $|Sy\rangle_i (i = 1, 2, 3, 4)$ denotes the ith syndrome

Then:

$$
\begin{aligned}
|\mathbf{00}\rangle &\to |\boldsymbol{E}_0\rangle = \boldsymbol{\alpha}|\mathbf{000}\rangle + \boldsymbol{\beta}|\mathbf{111}\rangle \\
|\mathbf{11}\rangle &\to |\boldsymbol{E}_1\rangle = \boldsymbol{\alpha}|\mathbf{100}\rangle + \boldsymbol{\beta}|\mathbf{011}\rangle \\
|\mathbf{10}\rangle &\to |\boldsymbol{E}_2\rangle = \boldsymbol{\alpha}|\mathbf{010}\rangle + \boldsymbol{\beta}|\mathbf{101}\rangle \\
|\mathbf{01}\rangle &\to |\boldsymbol{E}_3\rangle = \boldsymbol{\alpha}|\mathbf{001}\rangle + \boldsymbol{\beta}|\mathbf{110}\rangle
\end{aligned} \tag{30}
$$

In particular, each of the four Bell states can be written as a suitable quantum superposition of the four syndromes:

$$
\begin{aligned}
|\boldsymbol{\Phi}\rangle_\pm &= \frac{1}{\sqrt{2}}\left(|\boldsymbol{Sy}\rangle_0 \pm |\boldsymbol{Sy}\rangle_1\right) \\
|\boldsymbol{\Psi}\rangle_\pm &= \frac{1}{\sqrt{2}}\left(|\boldsymbol{Sy}\rangle_2 \pm |\boldsymbol{Sy}\rangle_3\right)
\end{aligned} \tag{31}
$$

In the case of study, we denote the quantum state of the Self as $|S\rangle = |0\rangle$, and the quantum state of the non-Self as $\left|S^\perp\right\rangle = |1\rangle$. The unknown quantum state of the quantum Self, $\left|\hat{S}\right\rangle$, is then:

$$
\left|\hat{S}\right\rangle = \alpha|S\rangle + \beta\left|S^\perp\right\rangle \tag{32}
$$

This is the first qubit in the encoded logical qubit $|\boldsymbol{\Psi}\rangle_L$.

The syndrome $|00\rangle$ says that there are no errors, neither in the first qubit, nor in the other two ones. In particular, this ensures that the relation between the Self and the non-Self is unchanged.

The syndrome $|11\rangle$ states that there is an error in the first qubit. In this case, there is a flip between the Self and the non-Self: $\alpha\left|S^\perp\right\rangle + \beta|S\rangle$.

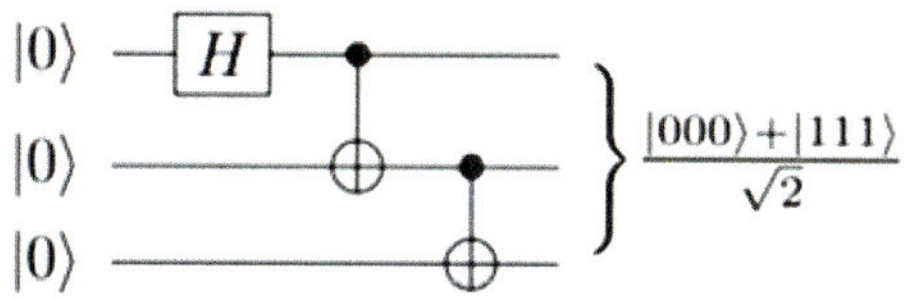

Fig. 4 Quantum circuit generating the GHZ state: an Hadamard gate H acting on the first qubit, followed by two CNOT gates: the first one acting on the first and second qubit, and the second one acting on the second and third qubit

The Bell state $|\boldsymbol{\Phi}\rangle_{\pm} = \frac{1}{\sqrt{2}}(|00\rangle \pm |11\rangle)$ then represents the quantum superposition of the two limit cases: (i) no error in the relation between the Self and the non-Self, and (ii) a flip between the Self and the non-Self.

It should be noted that if the Self and the non-Self are in a cat state $\left(\boldsymbol{\alpha} = \boldsymbol{\beta} = \frac{1}{\sqrt{2}}\right)$:

$$\left|\hat{S}\right\rangle = \frac{1}{\sqrt{2}}\left(|S\rangle + \left|S^{\perp}\right\rangle\right) \tag{33}$$

Then, the encoded logical qubit $|\Psi\rangle_L = \alpha|000\rangle + \beta|111\rangle$, for $\boldsymbol{\alpha} = \boldsymbol{\beta} = \frac{1}{\sqrt{2}}$ is the three-qubit maximally entangled state GHZ (Greenberger et al., 1989):

$$|\text{GHZ}\rangle = \frac{1}{\sqrt{2}}(|000\rangle + |111\rangle). \tag{34}$$

The GHZ state is generated by the quantum circuit with one Hadamard gate and two CNOT gates. See Fig. 4.

The Hadamard gate H acts on the first qubit: $\boldsymbol{H}|\mathbf{0}\rangle \rightarrow \frac{1}{\sqrt{2}}(|\mathbf{0}\rangle + |\mathbf{1}\rangle)$, then, a first CNOT gate entangles the cat state with the second qubit:

$$CNOT\left\{\begin{array}{c}\frac{1}{\sqrt{2}}(|\mathbf{0}\rangle + |\mathbf{1}\rangle)\\ |\mathbf{0}\rangle\end{array}\right. \rightarrow \frac{\mathbf{1}}{\sqrt{\mathbf{2}}}(|\mathbf{00}\rangle + |\mathbf{11}\rangle)$$

giving rise to the Bell state $|\Phi_{+} = \frac{1}{\sqrt{2}}(|00\rangle + |11\rangle)$, and finally, a second CNOT gate entangles the Bell state with the third qubit:

$$CNOT\left\{\begin{array}{c}\frac{1}{\sqrt{2}}(|\mathbf{00}\rangle + |\mathbf{11}\rangle)\\ |\mathbf{0}\rangle\end{array}\right. \rightarrow \frac{1}{\sqrt{2}}(|\mathbf{000}\rangle + |\mathbf{111}\rangle)$$

giving rise to the GHZ state.

3.2 Encoding the Quantum Self

As we already said in Sect. 2, the Self-cat state corresponds to the fixed point of the fuzzy complement operation, and it is completely unaffected by quantum bit-flip errors.

The resulting encoded logical qubit in this case is the GHZ state, which is a 3-qubit maximally entangled state.

In Sect. 2, we considered the universe U constituted by the fuzzy set "Self" and its complement "non-Self". Such a universe represents the closed quantum system of an isolated Self-qubit.

So where do the extra qubits needed to encode the quantum Self into the logical qubit come from? The answer is that the Self-qubit is not an isolated quantum system, but it can interact with its environment E, where the extra qubits are located.

The appearance or not of quantum errors depends mainly on the intensity of this interaction, as we will show below.

Let us consider the total quantum system:

$$S_{\text{Tot.}} = S + E \tag{35}$$

and let $\rho_S(0)$ and $\rho_E(0)$ be the density matrices of the system S and of the environment E, respectively, at the initial time $t = 0$. At the initial instant, the state of the total system is described by the tensor product:

$$\rho_{\text{Tot}}(0) = \rho_S(0) \otimes \rho_E(0). \tag{36}$$

If the interaction is weak, the time evolution is unitary, and the state is separable at every instant t:

$$\rho_{\text{Tot}}(t) = \rho_S(t) \otimes \rho_E(t). \tag{37}$$

That means that there is no entanglement between the system and the environment. If, on the other hand, the interaction is strong, the state is no longer separable. In the presence of entanglement, quantum errors can occur.

Note that when the quantum Self becomes entangled with its environment, it, like any entangled state, loses its identity: this is the loss of self-awareness.

In summary, the loss of self-awareness appears to be due to a very strong interaction of the quantum Self with its environment. Since this implies a large amount of quantum errors, a possible cure could be the application of a proper quantum error-correcting code.

Another tripartite maximally entangled state is the W state, the name "W" is for Dür et al. (2000), whose shape is:

$$W = \frac{|001\rangle + |010\rangle + |100\rangle}{\sqrt{3}} \tag{38}$$

The W state is the representative of one of the two non-biseparable classes of three-qubit states, the other being the GHZ state discussed above.

In a W-state, precisely one of the qubits is "on" (1), and the rest are "off" (0).

The Bell state $|\boldsymbol{\Psi}\rangle_+ = \frac{1}{\sqrt{2}}(|\mathbf{01}\rangle - |\mathbf{10}\rangle)$ is a W-state on two qubits, and $|\boldsymbol{\Psi}\rangle_- = \frac{1}{\sqrt{2}}(|\mathbf{01}\rangle - |\mathbf{10}\rangle)$ is a W-state up to phase.

W-states are used in quantum communication (Wang et al., 2007), in particular quantum teleportation (Bennett et al., 1993) and cryptography (Liu et al., 2011) protocols.

The GHZ and W states cannot be transformed into each other (Dür et al., 2000) by local quantum operations (LOCC) (Nielsen, 1999).

The GHZ state is fully separable after loss of one qubit. For this reason, a GHZ state is said a "three-party" entangled state, with tangle $\tau_{ABC}|\mathrm{GHZ}\rangle = 1$, where τ (the tangle) is a multipartite entanglement measure. Instead, if one of the three-qubits is lost in a W state, the remaining two qubits are entangled. Then, the W state is a two-party entanglement state, with tangle $\tau_{ABC}|W\rangle = 0$, but not a three-party entanglement state.

Indeed, W is the three-qubit state whose entanglement has the greatest robustness against the loss of a qubit (Dür et al., 2000).

While this property is very useful in quantum computing, for example, ensuring good storage properties of ensemble-based quantum memories, it can be a serious limitation when related to the quantum Self, as we will show in what follows.

Let us consider the logical quantum Self $\hat{S}$ in the three-qubits quantum error-correcting code:

$$|\Psi\rangle_{L(\hat{s})} = \alpha|S\rangle^{\otimes 3} + \beta\left|S^{\perp}\right\rangle^{\otimes 3}. \tag{39}$$

The three possible errors $|E_i\rangle(i = 1, 2, 3)$ in Eq. (30) can be rewritten as:

$$|E_i\rangle_{\hat{S}} = |E_i\rangle_S + |E_i\rangle_{S^{\perp}} \tag{40}$$

where:

$$|E_i\rangle_S = \{|001\rangle, |100\rangle, |010\rangle\} \tag{41}$$

and the $|E_i\rangle_{S^{\perp}}$ are given by $NOT|E_i\rangle_S$:

$$|E_i\rangle_{S^{\perp}} = \mathrm{NOT}|E_i\rangle = \{|110\rangle, |011\rangle, |101\rangle\}. \tag{42}$$

It should be noted that a suitable quantum superposition of the $|E_i\rangle_S$ gives the W state:

$$\frac{1}{\sqrt{3}}\sum_{i=1}^{3}|E_i\rangle_S \equiv W = \frac{|001\rangle + |100\rangle + |010\rangle}{\sqrt{3}} \tag{43}$$

which is a tripartite maximally entangled state.

Suppose now that the three possible errors $|E_i\rangle_S$ of $|S\rangle^{\otimes 3}$ merge into a W state before the quantum-error-correcting code is applied. Subsequently, even if a quantum measurement is performed on a qubit of the environment, the Self remains entangled with the environment and the cure cannot be further applied. Simply put, mental illness can be eased in this case, but mental health cannot be completely restored.

Instead, the sum of the three $|E_i\rangle_{S^\perp}$ regarding the non-Self is not an entangled state:

$$\sum_{i=1}^{3}|E_i\rangle_{S^\perp} = |110\rangle + |011\rangle + |101\rangle = |Sy\rangle_1 \otimes |0\rangle + \big(|Sy\rangle_2 + |Sy\rangle_3\big) \otimes |1\rangle \quad (44)$$

This can be interpreted as follows. The first term $|Sy\rangle_1 \otimes |0\rangle$ in Eq. (44) says that an error is detected in the first qubit (the Self), and there is a tensor product with one ancilla of the environment (there is no entanglement). The second term $\big(|Sy\rangle_2 + |Sy\rangle_3\big) \otimes |1\rangle \equiv |\Psi\rangle_+ \otimes |1\rangle$ says that the two ancilla are entangled to each other in the Bell state $|\Psi\rangle_+$, but the latter is not entangled with $\left|S^\perp\right\rangle \equiv |1\rangle$. The conclusion is that the non-Self does not get entangled with the environment, only the Self does.

This has a quite intuitive understanding; in fact, the non-Self is a part of the environment when the quantum Self is viewed as an open quantum system. More precisely, the non-Self can be viewed as the projection of the Self into the environment. See Fig. 5.

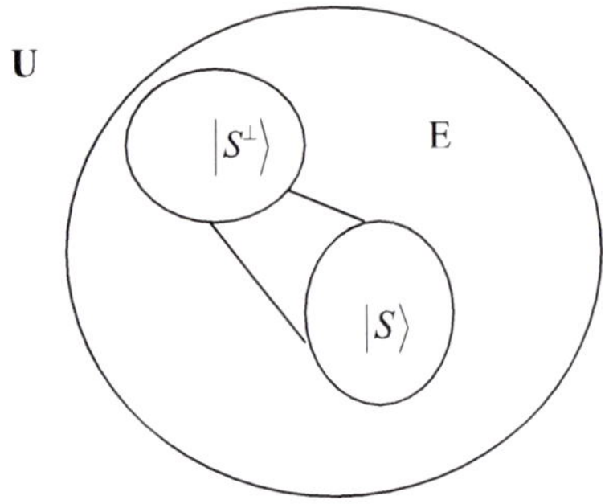

Fig. 5 Non-Self as a projection of the Self into the environment E

3.3 *Logical Pathologies of the Self*

A "fuzzy" Self, i.e., a Self that is characterized as an assertion $|-^{\alpha} S$ with a degree of assertion α, obeys the axiom of identity $S\Big|-^{|\alpha|^2} S$ with a degree of partial truth $|\alpha|^2 \in [0, 1]$.

This "fuzzy" Self is the potential origin of the quantum Self $\hat{S}_{\alpha\beta}$ that is obtained through the construction of the quantum logical conjunction $\alpha\&_{\beta}$ in the following way:

$$\frac{\big||-^{\alpha} S \quad |-^{\beta} S^{\perp}}{\Big|-\hat{S}_{\alpha\beta}}\ {}_{\alpha}\&_{\beta}\text{ - form} \tag{45}$$

where $\alpha\&_{\beta}$-form denotes the logical $\alpha\&_{\beta}$ formation rule, $\big|-^{\beta} S^{\perp}$ is the non-Self asserted with assertion degree β, and the explicit expression of the asserted quantum Self $\Big|-\hat{S}_{\alpha\beta}$ is: $|-S\,_{\alpha}\&_{\beta} S^{\perp}$.

We can say, in simple terms, that the statement $|-^{\alpha} S$ describes a semi-classical Self, from which a quantum Self $\Big|-\hat{S}_{\alpha\beta}$ can arise through appropriate logical processes.

Instead, the classical Self cannot be traced back to any assertion, because, as we have seen in Sect. 2, the classical identity axiom does not originate in the metalanguage.

In the latter case, the Self obeys the classical identity axiom, and a dichotomy arises between the Self and the "Other". We think this is the case with an autistic Self.

Let us now make a scheme in which the different types of identity axiom select the different types of Self:

(1) Classical identity axiom $S\big|-^{v=1} S$ Classical Self
(2) Quantum identity axiom $S_{\alpha}\&_{\beta} S^{\perp}\Big|-^{v-|\alpha|^2+|\beta|^2=1} S_{\alpha}\&_{\beta} S^{\perp}$ Quantum Self
(3) Semi-classical identity axiom $S\Big|-^{v=|\alpha|^2} S$ Semi-classical Self.

where v stands for truth value, and $|\alpha|^2$ and $|\beta|^2$ are partial truth values with $|\alpha|^2 + |\beta|^2 = 1$.

The alternation of the semi-classical identity axiom (2) and the quantum identity axiom (3) gives an alternation of conscious and unconscious Self in a normal Self. If only the quantum identity axiom is present, we are in presence of a schizophrenic Self. It is clear, as we discussed in Sect. 2.2, that the semi-classical Self and the identity axiom to which it obeys are both derived from a quantum metalanguage.

As was illustrated in Zizzi and Pregnolato (2012), a quantum logic describes creativity, the dream, and the unconscious, and, in pathological cases, schizophrenia.

The classical identity axiom belongs to classical (Aristotelian) logic as well as to non-classical logics, among which we recall, for example, intuitionistic logic

(Heyting, 1930) (with absence of the principle of double negation), quantum logic (Birkhoff, 1936) (with absence of distributivity between conjunction and disjunction), linear logic (Girard, 1987) (with finer control on structural rules), many-valued logic (Łukasiewicz, 1920) and fuzzy logic (Zadeh, 1968) (with partial truth-values), paraconsistent logic (Priest, 2002) (with refusal of the non-contradiction principle), basic logic (Sambin et al., 2000) (sub-structural, with the three principles of reflection, symmetry, and visibility). This means that the Self of an autistic subject, characterized by a classical identity axiom, can be a proposition of any logic apart from that derived from a quantum metalanguage. Therefore, the autistic Self can range from classical (Aristotelian) logic which is the most abstract and mechanical one, and it is in fact a logic not suitable for human reasoning, to any of the most sophisticated non-classical logics mentioned above, which are much more suitable for human reasoning even if not suitable to describe the unconscious thought.

In this scheme, the classical identity axiom is associated with autism due to the total lack of any quantum aspect, and therefore, total lack of openness toward the "Other". And it also explains the purely mechanical aspects of the autistic Self.

There is currently no cure for autism, only therapy. The possible cure would be to provide a quantum metalanguage to the autistic subject, but this process is currently unknown. However, in Zizzi (2013), it was shown that the statements of a quantum metalanguage are physically interpretable as coherent states (Glauber, 1963).

Hence, a possible cure for autism could be to provide coherent states to the classical Self through laser therapy to make it semi-classical. In fact, in a laser, light is emitted into a resonant mode, and that mode is highly coherent; thus, laser light is idealized as a coherent state. In support to our hypothesis, a low-level laser therapy in autism spectrum disorder (ASD) has recently been tested (Leisman et al., 2018). We will extend these ideas about autism and coherent states in a forthcoming paper.

Regarding schizophrenia, once confirmed, a definitive cure is not possible. According to our model, only prevention is possible, when the first symptoms make one suspect the disease, and this prevention consists in eliminating the quantum errors of the codified quantum Self.

In fact, when the disease is confirmed, it means that all errors of the semi-classical self $|-^{\alpha} S$ have formed the maximally entangled state W, as in Eq. (43), and therefore can no longer be corrected. Then, there is no longer any alternation between the semi-classical Self and the quantum Self, $\Big| - \hat{S}_{\alpha\beta}$, which therefore predominates, leading to schizophrenia.

In order to explain that more formally, we will show some features of the encoding of the unknown quantum Self into the logical qubit, and its possible errors, in terms of sequent calculus.

Sequent calculus is a deduction system for performing reasoning in first-order logic (and propositional logic). It was introduced by Gentzen (1935) initially for classic logic (LK) and then extended to intuitionist logic (LJ). It is now used for several different logics, among which sub-structural logics, like linear logic (Girard, 1987) and basic logic (Sambin et al., 2000). The latter has the property of visibility, that is, all active formulas A, B, C, … are independent from contexts Γ, Δ, Σ, Π...

The logic used in this paper is a quantum version of basic logic, and then, the property of visibility holds as well. However, when one considers an open quantum system as is the case when quantum errors are taken into account, there is no more control on the logical contexts, representing the environments. In this case, one should use the LK system instead of that of basic logic.

The encoding of the unknown quantum Self in Eq. (32) into the logical qubit in Eq. (39) can be described by the &-formation rule in the LK system:

$$\frac{|-^{\alpha} S, \Delta \qquad |-^{\beta} S^{\perp}, \Sigma}{|-S_{\alpha}\&_{\beta} S^{\perp}, \Delta, \Sigma} \ \&_{\beta} \text{ - form}_{\alpha} \tag{46}$$

where we have put the antecedent $\Gamma = 0$ everywhere as the sequents that we consider are assertions.

In the premises of Eq. (46), the Self S interacts with $\Delta = \{|0\rangle_2, |0\rangle_3\}$ and the non-Self $S^{\perp}$ interacts with $\Sigma = \{|1\rangle_2, |1\rangle_3\}$.

The consequence is that $S_{\alpha}\&_{\beta} S^{\perp}$ interacts with both Δ, Σ, and the logical Self-qubit is formed:

$$|-S_{\alpha}\&_{\beta} S^{\perp}, \Delta, \Sigma \Rightarrow^{I} |\Psi\rangle_L = \alpha|000\rangle + \beta|111\rangle \tag{47}$$

where *"I"* stands for "interpretation".

4 Self-Disorders (SDs)

Self-Disorders or ipseity disturbances (ipse is Latin for "self" or "itself) are non-psychotic alterations of subjective experience that include distortions of self-awareness (e.g., fading first-person perspective, waning sense of basic identity, depersonalization, and hyperreflectivity), autopsychic disorders (e.g., thought pressure or block, perceptualization of mental stream, and spatialization of thoughts), loss of common sense (e.g., perplexity), and existential alterations (e.g., solipsistic grandiosity). Such experiences define essential aspects of the clinical expressions of schizophrenia (Sass & Parnas, 2003; Raballo, 2012). The contemporary concept of Self-Disorders refers to a disturbed structure of phenomenal consciousness, that is, to a disturbed sense of the experiential or minimal self (Zahavi, 2014; Nordgaard et al., 2021). Thus, the "Self" that is proposed to be disturbed in schizophrenia is a very basic experiential sense of Self, that is, more specifically, the very first-personal structure of experience as measured by EASE (Examination of Anomalous Self-Experience) a semi-structured psychometric instrument for the assessment of anomalous self-experience, also referred to as "Self-Disorders" or "basic self-disturbances". EASE scores differentiate the schizophrenia spectrum disorders (including schizotypal personality disorder) from other conditions such as psychotic bipolar disorder and borderline personality disorder (Parnas et al., 2005, Parnas & Henriksen, 2014).

Autism spectrum disorder (ASD) and autism are both general terms for a group of complex disorders of brain development. These disorders are characterized, in varying degrees, by difficulties in social interaction, verbal and nonverbal communication, and repetitive behaviors. Evidence of Self-Disorders appears more complex in autism than in schizophrenia, due particularly to autistic language impairment (Lyons & Fitzgerald, 2013), nevertheless, Tordjman et al. proposed the relationships between schizophrenia and autism regarding body-self impairments leading to self-other differentiation impairments, a deficit of theory of mind and empathy, and their consequences on social communication (Tordjman et al., 2019). Photobiomodulation (PBM) involves the delivery of red and/or near-infrared light from a laser or a LED (light-emitting diode) to a particular part of the body. Some recent study examined the efficacy of low-level laser therapy, a form of photobiomodulation, for the treatment autistic spectrum disorder (Leisman et al., 2018; Hamblin, 2022). Positive results, both in children and in adults, have been obtained, suggesting a possible future non-pharmaceutical approach to treat ASD and supporting the theoretic evidence proposed in this paper.

The model we proposed also opens new perspectives to better understand Self-Disorders and social communication impairments in schizophrenia to develop preventive strategies based on quantum error-correcting codes.

5 Conclusions

In this paper, we have shown that the quantum Self originates in the quantum metalanguage and is physically interpreted as an unknown quantum state. It is the quantum superposition of the semi-classical Self and its negation.

The semi-classical Self is probabilistically equal to Itself (the "faded" self), and we have argued that the loss of self-awareness can only have a quantum description since the quantum Self is a "constructive object" that can be built and destroyed. We then described the semi-classical Self as a fuzzy set. We showed that the quantum Self-cat state is a fixed point that cannot be affected by bit-flip errors and thus is the best candidate to represent mental health. However, the Self-qubit is not an isolated quantum system and can interact with its environment, where extra qubits are found. If the interaction is strong, the state is no longer separable and quantum errors occur. We encoded the unknown quantum Self in a 3-qubit logic state and discussed possible errors, which can be detected and corrected by the 3-qubit quantum error-correcting code.

When the quantum Self becomes entangled with its environment, like any entangled state, it loses its identity: this is the loss of self-awareness. Conscious awareness disorders, or Self Disorders (DS), are a disrupted sense of Self and occur in patients with schizophrenia and autism spectrum disorder.

In the case of autism, the fact that quantum metalanguage statements are physically interpretable as coherent quantum states indicate a possible therapy. A laser light emitted in resonant mode can be idealized as a coherent state; thus, it is possible to propose a therapy for autism by providing coherent states to the classical Self. In the case of schizophrenia, we have proposed a possible preventive treatment through the application of quantum error-correcting codes. Further studies are needed to translate our conceptual and theoretical model into practical clinical applications.

References

Benioff, P. (1980). The computer as a physical system: A microscopic quantum mechanical Hamiltonian model of computers as represented by Turing machines. *Journal of Statistical Physics, 22*(5), 563–591.

Bennett, C. H., Brassard, G., Crépeau, C., Jozsa, R., Peres, A., & Wootters, W. K. (1993). Teleporting an unknown quantum state via dual classical and Einstein-Podolsky-Rosen channels. *Physical Review Letters, 70*(13), 1895–1899.

Birkhoff, G. V., & Neumann, J. (1936). The logic of quantum mechanics. *Annals of Mathematics, 37*, 823–843.

Bombin, H. (2010). Topological subsystem codes. *Physical Review A, 81*, 032301.

Dür, W., Vidal, G., & Cirac, J. I. (2000). Three qubits can be entangled in two inequivalent ways. *Physical Review A, 62*, 062314.

Deutsch, D. (1985). Quantum theory, the Church-Turing principle and the universal quantum computer. *Proceedings of the Royal Society of London, A, 400*, 97–117.

Dummett, M. A. E. (1977). *Elements of intuitionism*. Clarendon Press.

Feynman, R. (1982). Simulating physics with computers. *International Journal of Theoretical Physics, 21*(6/7), 467–488.

Fowler, A. G., Mariantoni, M., Martinis, J. M., & Cleland, A. N. (2012). Surface codes: Towards practical large-scale quantum computation. *Physical Review A, 86*, 032324.

Gentzen, G. (1935). Untersuchungen über das logische Schließen. *Mathematische Zeitschrift, 39*(176–210), 405–431.

Girard, J. Y. (1987). Linear logic. *Theoretical Computer Science, 50*, 1–102.

Glauber, R. J. (1963). Coherent and incoherent states of radiation field. *Physical Review, 131*, 2766–2788.

Greenberger, D. M., Horne, M. A., & Zeilinger, A. (1989). Going Beyond Bell's theorem. In M. Kafatos (Ed.) *Bell's theorem, quantum theory and conceptions of the universe* (pp. 69–72). Kluwer Academic.

Hamblin, M. R. (2022). Could photobiomodulation treat autism spectrum disororder? *Photobiomodulation, Photomedicine, and Laser Surgery, 40*(6), 367–369.

Heyting, A. (1930). Die formalen Regeln der intuitionistischen Logik, in three parts. *Sitzungsberichte der preussischen Akademie der Wissenschaften, 42–71*, 158–169.

Heyting, A. (1971). *Intuitionism, an introduction* (3rd ed.). North-Holland.

Johnston, A., & Malobou, C. (2013). *Self and emotional life: Philosophy, psychoanalysis, and neuroscience*. Columbia University Press. ISBN 9780231158312.

Kitaev, A. (2003). Fault tolerant quantum computation by anyons. *Annals of Physics, 303*, 2–30.

Leisman, G., Machado, C., Machado, Y., & Chinchilla-Acosta, M. (2018). Effects of low-level laser therapy in autism spectrum disorder. In M. Pokorski (Ed.), *Clinical medicine research. Advances in experimental medicine and biology* (Vol. 1116). Springer.

Liu, W., Wang, J. B., & Jiang, Z. T. (2011). An efficient protocol for the quantum private comparison of equality with W state. *Optics Communications, 284*(12), 3160–3163.

Łukasiewicz, J. (1920). O logice trójwartościowej (in Polish). *Ruch filozoficzny, 5*, 170–171. English translation. L. Borkowski (Ed.) (1970). On three-valued logic. In J. Łukasiewicz (Ed.), *Selected works* (pp. 87–88). Amsterdam.

Lyons, V., & Fitzgerald, M. (2013). Atypical sense of self in autism spectrum disorders: A neuro-cognitive perspective. In M. Fitzgerald (Ed.), *Recent advances in autism spectrum disorders* (Vol. I(31), pp. 749–770). IntechOpen.

Nielsen, M. A. (1999). Conditions for a class of entanglement transformations. *Physical Review Letters, 83*(2), 436–439. arXiv:quant-ph/9811053.

Nordgaard, J., Henriksen, M. G., Jansson, L., Handest, P., Møller, P., Rasmussen, A. R., Sandsten, K. E., Nilsson, L. S., Zandersen, M., Zahavi, D., & Parnas, J. (2021). Disordered selfhood in schizophrenia and the examination of anomalous self-experience: Accumulated evidence and experience. *Psychopathology, 54*, 275–281.

Parnas, J., & Henriksen, M. G. (2014). Disordered self in the schizophrenia spectrum: A clinical and research perspective. *Harvard Review of Psychiatry, 22*(5), 251–265.

Parnas, J., Moller, P., Kircher, T., Thalbitzer, J., Jansson, L., Handest, P., & Zahavi, D. (2005). EASE: Examination of anomalous self-experience. *Psychopathology, 38*(5), 236–258.

Priest, G. (2002). Paraconsistent logic. In D. Gabbay, & F. Guenthner (Eds.), *Handbook of philosophical logic* (Vol.6, 2nd ed., pp. 287–394). Kluwer.

Raballo, A. (2012). Self-disorders and the experiential core of schizophrenia spectrum vulnerability. *Psychiatria Danubina, 24*(Suppl 3), S303–S310.

Sambin, G., Battilotti, G., & Faggian, C. (2000). Basic logic: Reflection, symmetry, visibility. *The Journal of Symbolic Logic, 65*, 979–1013.

Sass, L. A., & Parnas, J. (2003). Schizophrenia, consciousness, and the self. *Schizophrenia Bulletin, 29*(3), 427–444.

Steane, A. M. (2006). A tutorial on quantum error correction. In G. Casati, D. L. Shepelyansky, & P. Zoller (Eds.), *Proceedings of the International School of Physics "Enrico Fermi", course CLXII, "Quantum Computers, Algorithms and Chaos"* (pp. 1–32). IOS Press.

Terhal, B. M. (2015). Quantum error correction for quantum memories. *Reviews of Modern Physics, 87*, 307.

Tordjman, S., Celume, M. P., Denis, L., Motillon, T., & Keromnes, G. (2019). Reframing schizophrenia and autism as bodily self-consciousness disorders leading to a deficit of theory of mind and empathy with social communication impairments. *Neuroscience and Biobehavioral Reviews, 103*, 401–413.

Troelstra, A. S. (1991). *History of constructivism in the twentieth century*. University of Amsterdam, ITLI Prepublication Series ML-91-05.

Wang, J., Quan, Z., & Tang, C. J. (2007). Quantum secure communication scheme with W state. *Communications in Theoretical Physics, 48*(4), 637–640.

Yao, X. C., Wang, T. X., Chen, H. Z., Gao, W. B., Fowler, A. G., Raussendorf, R., Chen, Z. B., Liu, N. L., Lu, C. Y., Deng, Y. J., Chen, Y. A., & Pan, J. W. (2012). Experimental demonstration of topological error correction. *Nature, 482*(7386), 489–494.

Zadeh, L. A. (1965). Fuzzy sets. *Information and Control, 8*, 338–353.

Zadeh, L. A. (1968). Fuzzy algorithms. *Information and Control, 5*, 94–102.

Zahavi, D. (2014). *Self and other: Exploring subjectivity, empathy, and shame*. Oxford University Press.

Zizzi, P. (2013). The uncertainty relation for quantum propositions. *International Journal of Theoretical Physics, 52*, 186.

Zizzi, P., & Pregnolato, M. (2012). Quantum logic of the unconscious and schizophrenia. *NeuroQuantology, 10*(3), 566–579.

Zizzi, P. (2010). From quantum metalanguage to the logic of qubits [PhD Thesis] arXiv:1003.5976

Complex Systems and Energy

Umberto Di Caprio and Mario R. Abram

Abstract Complexity emerges in many descriptions of systems. It is a concept that may assume many different forms, depending on the particular and specific characteristics of each system. Nonlinearity, multiplicity of variables, dissipation, stability, and hierarchy are some properties that impact directly on systems complexity. The investigation of a phenomenon is traditionally operated following two paradigms. The study of the theoretical models and the evaluation of their experimental properties are traditionally in accordance with TE (Theory, Experiment) paradigm. With the availability of computers the simulation became an important third component of TES (Theory, Experiment, Simulation) paradigm. Theory and simulation are powerful artificial methodologies and tools that enable to investigate the structure of systems and processes, but the execution of experiments remains the only way to correctly evaluate the behavior of real systems. Is complexity in the descriptions of phenomena? Or is it a property of the phenomena? For a system this dichotomy emerges when it is necessary to understand the role of energy and energy transformations. In particular the involved powers cover a crucial role because they impact directly on the interactions among the systems and processes.

1 Introduction

The need to understand and to control phenomena, that around us impact greatly on our lives, motivated the developments of investigation capabilities and methodologies. The increasing of knowledge about our environment and the interactions between the many systems reached the rank of scientific method that is traditionally based on the TE (Theory, Experiment) paradigm.

U. Di Caprio · M. R. Abram (✉)
AIRS (Associazione Italiana per la Ricerca sui Sistemi), Milano, Italy
e-mail: mario.abram@alice.it

U. Di Caprio
e-mail: umbertodicaprio@tiscali.it

G. Minati and M. Pietronilla Penna (eds.), *Multiple Systems*, Contributions to Management Science, https://doi.org/10.1007/978-3-031-44685-6_14

The availability of computer technologies accelerated the development of simulation methodologies. Then simulation became a crucial resource because it enabled to develop "simulation experiments" overcoming many very critical limitations of theory and experiment. Simulation then became a third "environment" in which many models that are too hard to manage only by theory or are very difficult to investigate only by experiments may find an useful environment for "experimenting theory" and for "speculating about experiments".

Then the TES (Theory, Experiment, Simulation) paradigm became quickly the new conceptual reference in sciences developments and engineering realizations. These aspects gained great value when the investigation and design activities of complex engineering realizations called for the effective development, implementation, and testing of adequate control functions. These are very critical aspects that emerge during the study and development of "complex systems engineering".

In this paper, two possible investigation paradigms, as TE (Theory, Experiment) and TES (Theory, Experiment, Simulation), are recalled. In particular the characterization of the context in which the variables are involved is described (Sect. 2). When multiple systems interact, as in engineering realizations, it is possible to consider processes and to evaluate the effective characterization of the experimental operating ranges (Sect. 3). The role of energy may be correctly evaluated only in experimental phases, but the energy interactions cover a central role to correlate the natural (experiment) and artificial (theory and simulation) contexts (Sect. 4). In particular energy dynamics is important because it gives an effective evaluation of dissipation and of the system stability region (Sect. 5). In TE and TES paradigms the emergence of complexity and its different interpretations and implications are shortly investigated (Sect. 6). Some remarks and open problems show the need for further investigations (Sect. 7). Finally some conclusive considerations close the paper (Sect. 8).

2 Reference Paradigms

Investigating the behavior of natural phenomena is an human activity that evolved during the ages. In particular the experience, gained when working directly on the phenomena, found its natural context into the experimental activities. The need to understand the properties of phenomena and to formalize their description led to develop many theoretical activities that gained great advantages from the application of mathematics and physics in all their branches.

Theories were developed in order to explain the results of experiments. But in addition the experiments constitute the real check of theoretical ideas and new experimental results become the stimulus for further refinement of theories. This approach is represented by the TE (Theory, Experiment) paradigm (Fig. 1a); it was for long time the kernel of scientific developments in many disciplines.

When analog and digital computers became available, the simulation methodologies gained space and credibility. Now simulation is considered an indispensable step for the advancement of research and for design and production activities.

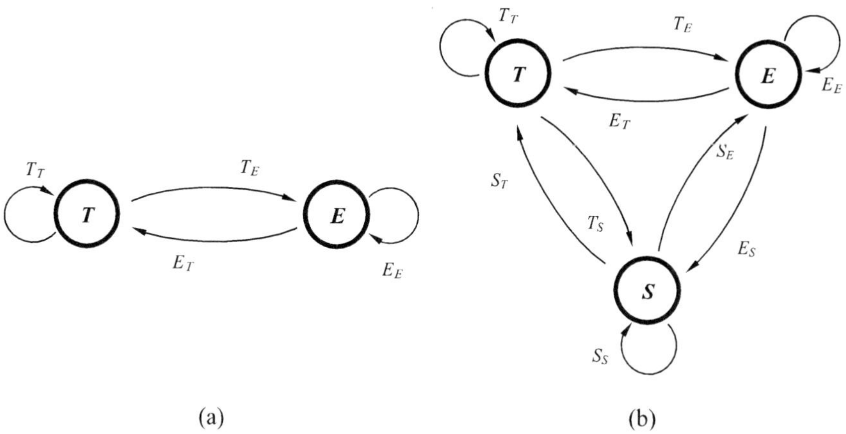

Fig. 1 Two paradigms. **a** TE: theory (T) and experiment (E). **b** TES: theory (T) and experiment (E) with simulation (S)

As a consequence a new paradigm, that we call TES (Theory, Experiment, Simulation) (Fig. 1b), emerged naturally. Now TES is the paradigm that inspires the activities for research, design, and testing that may gain great advantages from the interactions of theory, experiment, and simulation. It is useful for the development and production of a great variety of realizations and services.

The TE and TES paradigms constitute the conceptual maps of the processes involved in developing knowledge. In particular TES is the evolution of TE paradigm. These elements and their interactions constitute the kernel of every investigation criteria useful to develop uncountable scientific and technical applications. In Fig. 1 an arrow E_T visualizes how the information available from the experiment E may be used as an input for the theory T. So Fig. 1 shows the possible "operating arrows" available for TE and TES paradigms.

The nature of systems and their description define the numerical computation ranges of each representation status. For each variable V it is possible to define the operating ranges related to TE and TES representations as follows (Fig. 2).

- For a theoretical representation the *theoretical range* R_T of a variable V may be considered; the field of definition of the variable V for the models is usually given by $R_T = (T_m, T_M)$ in which T_m is the minimum theoretical value of V and T_M is the maximum theoretical value of V. In particular the conditions $T_m = -\infty$ and $T_M = +\infty$ are acceptable, depending on the mathematical structure of the model.
- In simulation model the *simulation range* R_S is considered. The simulation range is settled by the maximum representable numerical value of the simulation, that is the computing range of the system executing simulation. The range of a simulation model is $R_S = (S_m, S_M)$, in which S_m is the minimum value of V and S_M is the maximum value of V, of the representable numerical values for a specific computer system.

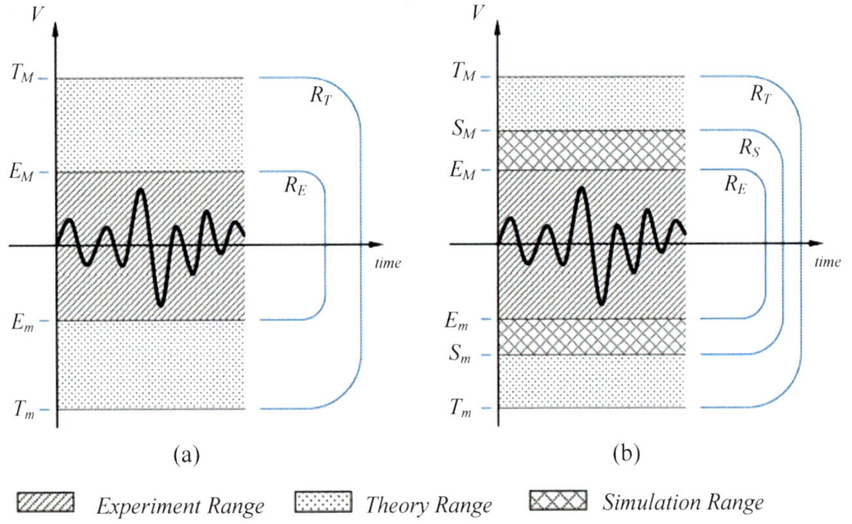

Fig. 2 Possible ranges in TE and TES paradigms for a variable V evolving in time. **a** The ranges for TE: theory (T); and experiment (E). **b** The ranges for TES: theory (T) and experiment (E) mediated by simulation (S)

- For the natural and artificial systems the *experiment range* R_E of a variable V is limited within the range of measurable values. Then the range for experiment is $R_E = (E_m, E_M)$, in which E_m is the minimum value and E_M is the maximum value reachable by the variable V during the execution of the experiment.

For a variable V the range of each representation has a general property that shows the real context in which model, simulation, and experiment may be considered:

$$R_E \subset R_S \subset R_T.$$

These relations evidence how the experiments are submitted to specific and strict physical bonds in their validity ranges while, in general, simulation models and theoretical models may operate without any bonds.

Considering TE and TES paradigms, a qualitative picture of the ranges hierarchy for a variable V is shown in Fig. 2. For TE paradigm (Fig. 2a), we have:

$$T_m < E_m < V < E_M < T_M,$$

while for TES paradigm (Fig. 2b) we may write:

$$T_m < S_m < E_m < V < E_M < S_M < T_M.$$

3 Processes

In engineering applications we prefer to use the name *process* when the interactions between systems are involved. These interactions may directly and dynamically impact on the functioning limits of operating and safety conditions.

Processes are originated by interacting systems that mutually exchange information, energy, and power. In processes the range value for a variable is dynamic and is directly influenced by the interacting systems. In fact the experimental range of each system may be dynamically adapted to guarantee the respect of all the process constraints.

It is important then to evaluate the operating range of all the processes and to design the operating range of each system according to a complete and global process design.

Working on real-world processes, for the range R_P of a process variable V_P, it may be convenient to consider the *process operating range* R_{PO} and the *process safety range* R_{PS}. The process operating range is defined as $R_{PO} = (PO_m, PO_M)$ in which PO_m is the minimum value and PO_M is the maximum value reachable by the processes variable V_P in operating conditions. Similarly the process safety range is given by $R_{PS} = (PS_m, PS_M)$ in which PS_m is the minimum value and PS_M is the maximum value reachable by the processes variable V_P in safety conditions. For definition we consider $R_P = R_{PS}$ and consequently $P_m = PS_m$ and $P_M = PS_M$.

Then for the process ranges, we may write the conditions (Fig. 3):

$$R_{PO} \subset R_{PS} = R_P \subset R_S \subset R_T.$$

Usually the operating range R_{PO} contains the range values of normal operations that may be reached with continuity in time, while the safety range R_{PS} may exceed the operating range for a limited and pre-definite time interval. They are the values that define the limits for normal operation and for the operation in dangerous or emergency conditions. Usually R_{PS} defines the limit conditions that, if they are reached or exceeded, may activate the starting of block procedures with the goal to stop that part of the process.

A qualitative picture of the ranges hierarchy for a process variable V_P is shown in Fig. 3. In particular for TE paradigm (Fig. 3a), we have:

$$T_m < P_m = PS_m < PO_m < V_P < PO_M < PS_M = P_M < T_M,$$

while for TES paradigm (Fig. 3b), we may write:

$$T_m < S_m < P_m = PS_m < PO_m < V_P < PO_M < PS_M = P_M < S_M < T_M.$$

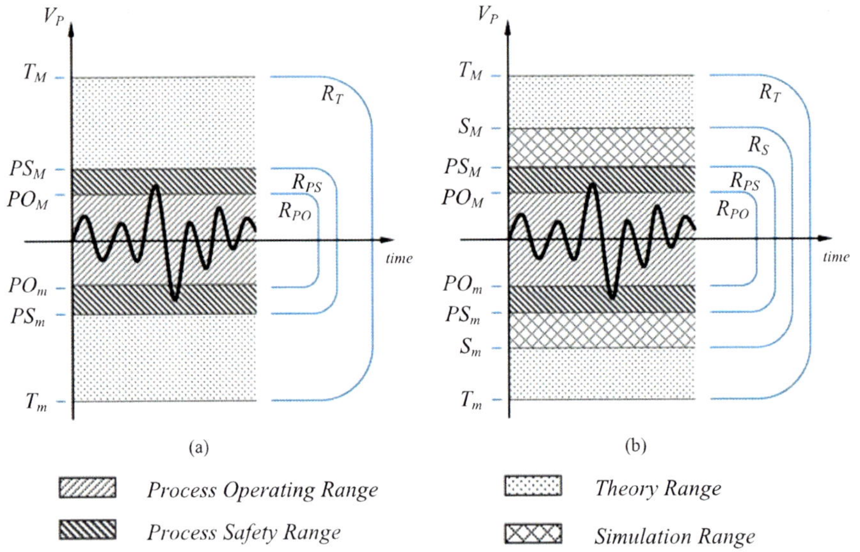

Fig. 3 Possible ranges in TE and TES paradigms for a process variable V_P evolving in time. **a** The ranges for TE: theory *T*; process operating (PO) and process safety (PS). **b** The ranges for TES: theory *T* and process operating (PO) and process safety (PS) mediated by simulation (*S*)

4 The Role of Energy

As we saw, theory and simulation are now fundamental tools in design and development activities. They are human activities that find their conceptual basis in modeling and in studying the dynamics of time evolution of models.

But the experiment phase remains the kernel of each test and verification activity. Then the experiment is indispensable; it is a fundamental activity because it enables to verify and validate the correctness of theoretical and simulation models.

The energy transformations may be modeled and used in theory and simulation, but only in experiments all the energy transformations and the consequent power balances are real because they are the natural components of the phenomena. Then the dynamics of energy in experiments is natural because energy transformation are intrinsic properties of real systems.

On the contrary modeling and consequently simulation, working on theories and models, treat energy as the result of models behavior. This appears evident when the models describe only a part of reality and then some phenomena are not described because they are considered marginal or insufficiently known. This appears more evident when we attempt to build a model of energy and power transformations.

Furthermore the energy in real systems is finite. For this reason the dynamical system describing a phenomenon must consider nonlinearities and finite range of the involved variables in order to assure the correct modeling of finite energy and power transformations.

The bounded values assumed by variables in the experiment implicate that the ranges of the involved variables necessarily must be bounded. Then all the models must have bounded variables, as in real world, in order to assure meaningful values for energies and powers. The correct energy and power balances become then a measure of the goodness and affordability of a model.

These requirements are a must if we have the goal to reach the maximum coherence between the hypothesis and calculations in theoretical and simulation models and the results of experiments. In general this becomes a very critical point that must be guaranteed in designing and operating realistic plant simulators.

5 Dissipation and Stability

The concepts of dissipation and stability are strictly connected to energy. The dissipative systems give a realistic description for a great part of natural phenomena. All the previous considerations acquire increasing importance, and they become critical to evaluate correctly the role of dissipation in a systems.

Then the energy values, with the help of constant energy surfaces, are useful to evaluate the stability region of the system. This information helps to determine if a process may be stable or unstable and it is essential for the design of appropriate control actions.

The study and determination of the stability region of a system may be helpful to visualize and to explain specific properties of the system dynamics. The picture of stability region may be of great help to explain, qualitatively and quantitatively, the dynamic behavior of the system. The evaluation of the correct value of energy defines the properties of stability and instability areas and characterizes their bounds and their shapes.

These considerations acquire great importance when operating with dissipative systems. The modeling of energy transformations and the consequent power balances call for a specific attention, and they must be a necessary target for theory and simulation. Technically we may say that the models used in theory and simulation, if they must describe faithfully the real world, must consider dissipation and all the tools necessary to guarantee that the energy transformations and consequently of power balances are correctly modeled.

The attempt to evaluate the stability of a system (Hahn, 1967) stimulated the search of theoretical and experimental tools with the goal to bypass the difficult theoretical investigations that need the explicit solutions of the involved differential equations and the execution of complex experimental procedures. An example is given by the Lyapunov second method (Lyapunov, 1892) by which the stability properties of a dynamical system can be determined without the need to solve the associated differential equations.

In a similar way, simulation attempts to determine the evolution in time of a dynamical model without the need to explicitly solve the associated theoretical differential equations.

6 Complex Systems

Complexity is a term that collects all the difficulties we may encounter to understand phenomena, to build theoretical models, to search and to find all the possible interactions with the experimental activities. Then synthetically complexity expresses the difficulties to find and to determine the structure, the values and the functional descriptions of a phenomenon.

Some substantial questions may emerge. Is complexity a part of the theoretical and simulation descriptions of a phenomenon? Or is complexity an intrinsic property of the phenomenon? Is complexity a property of the mode, instruments, and methods used for describing a phenomenon?

Complexity emerges in many descriptions of systems. It is a concept that can assume many different forms, depending on the particular and specific characteristics of each system (Nicolis & Prigogine, 1989). Nonlinearity, multiplicity of variables, dissipation, stability, and hierarchy are some properties that impact directly on system complexity.

In experiment complexity lies in the phenomenon and may express the difficulty to identify and to measure the values of chosen variables.

For theoretical and simulation works we may attempt to evaluate complexity considering all the involved tools. Using more effective simulation tools and instruments, it should be possible to simplify theory and simulation and so the complexity of models and theories may be managed (Torresy et al., 2021).

Complexity is a property strictly connected to the knowledge and to the tools we actually use. The complex systems are characterized by dynamics that present non-predictable simple evolution for the presence of nonlinearity, resonance, or structural changes (Debarsy et al., 2017; Mainzer, 1997).

An additional aspect of complexity lies in the difficulty to describe and to use the information available from the states and relations of TE and TES paradigms. In particular for TE paradigm we have two states (T, E) and four relations (Fig. 1a). Similarly for TES paradigm we have three states (T, E, S) and nine relations (Fig. 1b). The relations represented in the graphs of Fig. 1 may be shown in matrix form, and we may consider the following incidence matrices:

$$M_{\mathrm{TE}} = \begin{bmatrix} T_{\mathrm{T}} & T_{\mathrm{E}} \\ E_{\mathrm{T}} & E_{\mathrm{E}} \end{bmatrix} \quad M_{\mathrm{TES}} = \begin{bmatrix} T_{\mathrm{T}} & T_{\mathrm{E}} & T_{\mathrm{S}} \\ E_{\mathrm{T}} & E_{\mathrm{E}} & E_{\mathrm{S}} \\ S_{\mathrm{T}} & S_{\mathrm{E}} & S_{\mathrm{S}} \end{bmatrix}$$

The arrows in Fig. 1 may be seen as the paths to the development of new knowledge. In particular the number of states and the number of relations may be seen as a first measure of the complexity in TE and TES paradigms.

The role of safety conditions and their implementation contribute to increase the complexity of systems and processes. In particular the control strategies must assure the processes safety in all the dynamic conditions. This means to operate in accordance with programmable conditions by which it is possible to modify and to adapt the operating and safety ranges of a process. Safety bounds are imposed by the conditions for interactions between the many systems of a process. In addition the complexity in engineering applications may rise, due to interactions between systems and processes and the managing of the high level operating and safety control functions. A further aspect of complexity may emerge considering the coexistence and the management of all these requirements (Debarsy et al., 2017; Liu & Barabási, 2016).

7 Remarks

Some key points appear important in order to identify the goals, to define the priorities for the activities, and to plan their realization. Some considerations help to go deeper and to show some interesting aspects from which useful suggestions for future improvements may emerge.

- *Energy*. The experiment remains the real context necessary to evaluate the energy transformations and the true balance of the involved energies and powers.
- *Finite energy*. The correct use of energy, and consequently of power, requires that only finite process greatness may be used for their description. Our modeling capability then needs to model correctly the transformations and the evolution in time of the finite values of energy and power.
- *Role of the experiment*. The role of energy, and then of power, can be evaluated correctly only with the help of experiments. At the end the experiment remains the only way to evaluate the role and the impact of energy transformations, then to understand the true dynamics of the involved powers.
- *Interactions*. The complexity of involved systems and processes enlarge the spectrum of behaviors and evolutions of dynamical modes. In processes the role of interactions between systems is a central problem and impacts directly on energy and powers transformations.
- *Environment*. The interactions between systems, processes, and their own environments appear a very critical subject. This is present when the exchanges of energy and the flows of the involved powers cover a crucial role.
- *Modeling*. In the experiments energy is present naturally, but in our models energy and power are present and are useful only if we are able to consider and to evaluate correctly and coherently their transformations.
- *Stability*. The energy surfaces enable to build a more general picture of the stability regions of the system. It is then possible to give a qualitative description of the system dynamics. Stability regions of a system are important in order to choice the working points for that system.
- *Control*. By means of control actions the dynamic evolution of a system is maintained within desired and predefined stability regions and contemporary its distance from instability regions is assured.

- *Engineering applications.* Stability and control are the key concepts for the realization of effective engineering applications composed by many interacting processes. One of the main goals is to guarantee safety, that is to assure the correct and secure operation of the processes, even in emergency conditions.
- *Complexity.* Complexity lies in modeling or is it a property of real systems? This means perhaps that reality is described correctly by the tools we are using, or is it necessary that new tools must be searched, developed, and used? It is a multifaceted concept that acquires new meaning from the interactions between systems and processes in evolving safety conditions.

8 Conclusions

The acknowledgment of the TE (Theory, Experiment) and TES (Theory, Experiment, Simulation) paradigms constitute the background conceptual reference on which the human activities may find their roots for the discovery and developments of new knowledge.

Taking into account the possible technologies evolution, the application of TES paradigm may follow two main directions: (1) the improvement and refinement of known methods and tools; (2) the development of new methods and the implementation of new tools as a consequence of the advancement of knowledge.

Complexity is relative to the actual knowledge and effective availability of theoretical, simulation, and experimental methodologies and tools. It is interesting to evaluate our ability to understand the limits of actual knowledge and of available tools in order to plan further improvements. In this approach the meaning of complexity may be revised and modified. This may lead to a more effective measurement and management of the many different aspects of complexity.

References

Debarsy, N., Cordier, S., Ertur, C., Nemo, F., Nourrit-Lucas, D., & Poisson, G. (Eds.). (2017). *Understanding interactions in complex systems: Toward a science of interaction.* Cambridge Scholars Publishing.

Hahn, H. (1967). *Stability of motion.* Springer-Verlag.

Liu, Y.-Y., & Barabási, A.-L. (2016). Control principles of complex systems. *Reviews of Modern Physics, 88*, 1–58.

Lyapunov, M. A. (1892). The general problem of the stability of motion. *Comm. Soc. Math.*, Kharkov. (A translation of the original paper published in Russia is available as a commemorative issue 1992). *International Journal of Control, 55*, 531–773.

Mainzer, K. (1997). *Thinking in complexity. The complex dynamics of matter, mind, and mankind* (3rd ed.). Springer-Verlag.

Nicolis, G., & Prigogine, I. (1989). *Exploring complexity. An introduction.* Freeman.

Torresy, L., Blevinsz, A. S., Bassettz, D., & Eliassi-Rady, T. (2021). The why, how, and when of representations for complex systems. *SIAM Review, 63*(3), 435–485.

A Cybernetic Perspective of Agent–Environment Relations: From Interactions to Meanings

Andrea Roli and Michele Braccini

Abstract We discuss the implications of a cybernetic perspective of the notion of *interaction* between a system and its environment. Starting from the classical artificial intelligence model of a robot as a system composed of sensors, actuators, and a control core, we emphasize the crucial importance of the interface between the system and its environment, i.e., its sensors and actuators. These components are not just interaction devices, but they are carriers of the semantic world of the system and are tightly linked to agent's goals and the characteristics of the environment that are relevant to the agent. The interactions between an agent and its environment are in general dynamic and open, specific to the contingent situation and dependent on their unique history. The capability of creating their own sensors and actuators (as it happens in natural evolution) enables agents to make sense of the world by creating their own semantic categories. We can envision analogous processes also in artificial systems, yet with some limitations. Finally, we illustrate and analyze some recent results on structural adaptation of artificial agents that (loosely) resembles the evolution of sensors in living organisms, and we discuss further steps toward some degree of epistemic autonomy in artificial systems.

1 Introduction

Before the introduction of the term *cybernetics* as the "control and communication in the animal and the machine" (Wiener, 1948), the subject was named "circular causal and feedback mechanisms in biological and social systems" (Von Foerster,

A. Roli (✉) · M. Braccini
Department of Computer Science and Engineering, Campus of Cesena, Università di Bologna, Via dell'Università 50, 47522 Cesena, Italy
e-mail: andrea.roli@unibo.it

M. Braccini
e-mail: m.braccini@unibo.it

A. Roli
European Centre for Living Technology, Ca' Bottacin, Dorsoduro 3911, Calle Crosera, 30123 Venezia, Italy

G. Minati and M. Pietronilla Penna (eds.), *Multiple Systems*, Contributions to Management Science, https://doi.org/10.1007/978-3-031-44685-6_15

2003) to sharply emphasize the role of circular causality in complex natural and artificial systems.[1] In this work, we take a cybernetic perspective to review the fundamental role of interactions between a system—either natural or artificial—and its environment. As neatly stated by Minati (2022), interactions, rather than being simple descriptive elements of system structure, can be seen as generators of complex systems. We usually denote by interaction a process that induces a behavioral dependence between two or more things (a special case is a self-interaction, mediated by the environment). A wide spectrum of properties and possible instances characterize interactions. Minati (2022) advocates for a principled and formal understanding of the notion of interaction that makes it possible to fully appreciate its role in generating multiple dynamics and complexity in systems. In particular, when we focus on the interaction between an *agent*—be it a robot or an organism—and the environment in which it is situated, the interaction is not just a perturbation, but it may carry a meaning, which depends on the cognitive expectations of the agent. The behavior of robots is a case in point, as it clearly exemplifies the emerging nature of behavior, which arises from the intertwined relations among the controller, the physical properties of the robot—including sensors and actuators—and the environment.

Starting from this observation, in this contribution, we discuss the generative role of interactions both in artificial and natural autonomous systems, and we emphasize the role of *affordances* in making interactions open and often with a non-predictable outcome. A further step of the argument is the acknowledgement that the interactions between an agent and the environment are connoted by the creation of meaning. For example, sensing organs emerge in organisms as they "make sense" of the outer world, conveying to the organism relevant information for its survival. We conclude the discussion by illustrating an ongoing approach toward the design of artificial systems (i.e., robots) endowed with the capability of shaping their own sensors and actuators, in the perspective of the design of autonomous epistemic devices.

2 Robot Behavior and the Sensory-Motor Loop

The common understanding of robot behavior is generally expressed in terms of a process emerging from the interplay among robot controller, robot body, and environment (Pfeifer & Scheier, 1999).[2] The controller can be a program running on a microprocessor, a circuit, or any physical device that transforms the inputs coming from the environment to actions executed by the robot. Emergence is not an all-or-nothing property, but it can rather happen at varying degrees: from simple behaviors

[1] Von Foerster trenchantly writes that cybernetics is "the only discipline that has given us a rigorous treatment of circular causality". In (Von Foerster, 2003), page 229.

[2] Here we are considering systems composed of one or more autonomous robots, discarding special cases such as teleoperated robots or industrial robots repeating the same sequences of actions.

arising from the interaction between a simple pure reactive system and its environment (Simon, 2019), to the self-organized behavior of swarms of robots. Let us first focus on the single-robot case, possibly characterized by non-trivial dynamical behavior (Nehmzow, 2008): knowing the controller is not enough to predict robot's behavior and, in general, the same controller can give rise to different behaviors depending on the body of the robot and the environment. Indeed, it is possible to achieve different behaviors by solely changing the environmental niche of the robot (Pfeifer & Scheier, 1999). For example, it is possible to tune the gait of a quadruped robot by changing the characteristics of the terrain. In dynamical systems terms, the same dynamical core can produce different dynamics depending on its coupling with the environment. In the case of swarms of robots (Brambilla et al., 2013), this emergent phenomenon is even more evident, as the behavior of the entire swarm is also a function of the interactions among robots. The interactions that give origin to the behavior of a single robot or a robot swarm may be regular or extemporaneous, with fixed or varying dynamics, and may appear and disappear depending on the situation.

Sensors and actuators mediate the interactions between a robot and its environment. The former ones are the devices that capture some physical features of the world and produce related data to the controller. Conversely, the latter ones are the devices that enable the robot to act on the environment. The relation between a robot and the environment (including also other robots) is circular: the percepts captured by the sensors are used by the controller to elaborate and send signals to the actuators, thus enabling the robot to act. Robot's actions, in turn, may produce new inputs to the sensors. This circular causal loop is called the sensory-motor loop (Pfeifer & Scheier, 1999) and is at the core of robot–environment interactions. The sensory-motor loop is the essence of robot behavior, from simple control feedback to complex and structured sequences of actions. For example, a simple feedback can be used to obtain phototaxis or collision avoidance behaviors (Braitenberg, 1986), while a more elaborated exploitation of the sensory-motor coordination enables manipulation capabilities (Metta & Fitzpatrick, 2003). Two different kinds of interactions take place in this scenario: (i) sensor–environment and actuator–environment relations at a first level, and (ii) at a higher level the sensory-motor loop, which plays upon the first interaction level.

All the above considerations can be translated, *mutatis mutandis*, to the case of living organisms. The comparison of the properties of natural and artificial agents is out of the scope of this paper.[3] However, it is important to anticipate here a point that will be discussed in Sect. 3: natural organisms have evolved their own way for sensing the environment and acting on it, and therefore, evolution has shaped their interactions in such a way that those features of the physical world that are relevant to the organisms can be extracted by sensing organs and the actions that can be relevant for surviving are enabled by features of the phenotype (e.g., arms and wings). Finally, *en passant* we observe that, being the behavior an emergent process involving control,

[3] The interested reader can refer to (Roli & Kauffman, 2020; Roli & Kauffman, 2022; Roli et al., 2022) for our perspective on this subject.

body, and environment, huge implications for neuroscience arise: knowing a neural circuit does not mean to know the function that the organism performs when this circuit is activated, and the interactions between an organism and the physical world induce changes in the brain.

In this work, we mainly concentrate on the first kind of interaction, i.e., sensor–environment and actuator–environment relations, which is central for the openness of complex systems. This kind of interaction originates from *affordances*, which are opportunities or impediments on the route of an agent to attain its goals. In the next section, we elaborate on the concept of affordance and its relation to meaning making in agents.

3 Affordances and the Emergence of Meaning

The term "affordance" has been introduced by Gibson (1966) in the context of ecological psychology to express the property of objects to afford observers possible actions. The word has been later adopted in several diverse fields, such as biosemiotics (Campbell et al., 2019) and robotics (Jamone et al., 2016). In general, an affordance can be defined as "a possible use of X by an organism to accomplish Y". These accomplishments can also occur through natural selection. Crucially, affordances are not independent features of the environment (Walsh, 2015), as a change can be neutral or not depending on the conditions of the organism, its goals, and its repertoire of actions. Affordances refer to what the environment offers to an agent. They may appear as either opportunities or obstacles on the agent's path to achieve a goal. A recent philosophical account (Heras-Escribano, 2019) emphasizes the relation between an organism and its perceived environment [its *umwelt*, according to von Uexküll terminology (von Uexküll, 2010)], stating that affordances guide and constrain the behavior of organisms, precluding or allowing them some actions, showing them what they can and cannot do. A step, for instance, affords us the action of climbing; a locked door prevents us from entering. "Affordances fill our world with meaning: organisms do not live in an inert environment, but are surrounded by promises and threats".[4]

Two aspects of affordance in robotics emphasized by Jamone and co-authors (2016) are particularly relevant for our discussion: (a) affordances are not properties belonging to the environment alone, but they depend on the actual relation between the environment and the current sensing and actuating abilities of the agent; and (b) affordance perception suggests action possibilities to the agent through the activation of sensory-motor patterns, and it also provides a mean to predict the consequences of actions. In current robotic research and applications, robot are equipped ab initio with sensors and actuators chosen by their designers. Therefore, the designers themselves decide what are the environmental features that are relevant for the robot: almost nothing is left open to be autonomously discovered and used by the robot. This is of

[4] Quoted from (Roli et al., 2022).

course a practical requirement, not just aimed at deploying robots able to perform given tasks, but also to provide some safety properties. However, the capability of adapting to changing environments and conditions, exploiting new ways for sensing and acting, might be useful in some robotic applications, such as operations in hostile environments. Moreover, the understanding of the role of affordances in "sense" and "act" functions in robots is critical for a well-founded assessment of human–machine interactions.

Contrasting robots with natural organisms makes it possible to elucidate the role of sensors and actuators in the creation of meaning in agents, both natural and artificial. The starting point is the discussion of the notion of affordances in biosemiotics (Hoffmeyer, 2008; Kull, 1999; von Uexküll, 2010). Hoffmeyer neatly illustrates the fundamental importance of "situatedness" in organisms and cites Brooks (1990) in observing that organisms use the world as their own best model: organisms are characterized by semiotically situated interactions, i.e., every interaction between the embodied agent and the environment carries a meaning that is specific to the agent. For natural organisms, sensing organs and those organs used to act in the physical space ("acting organs") have evolved through heritable variations and selection: evolution has shaped those organs that capture *semantic information* or enable the organism to effectively act in the environment and take advantage of those actions, i.e., organs that make it possible for the organism to act meaningfully with respect to its goals (Roli & Kauffman, 2020). Besides evolution, also learning enables organisms to develop new kinds of interactions with the environment. The problem is "learning to control some intrinsic dynamics so as to achieve a goal" (Hoffmeyer, 2008). This is actual *semiotic emergence*: the emergence of higher-level interaction patterns is the result of semiotic—and not just physical—interactions between agents and environment.

In both contexts, biological and artificial, we can say that while meaning is contingent, and therefore, adaptive through learning, since it is dependent on affordances arising from the specific interactions occurring between environment and organism, the degree of interpretive capacity is characteristic of the organism and is the result of evolution. Exemplifying this concept in terms of dynamical systems and linguistics, it can be said that the meaning of a signal derived from the environment depends on the specific dynamics that is currently taking place in the organism, and thus on the grammar with which the organism is interpreting the syntagmas/signals that reach it; while the totality of the dynamics expressible by the organism, which can be represented by the landscape of attractors (Roli & Braccini, 2018), represents the set of grammars that an organism is able to express/interpret.

Summarizing, our focus is on a specific class of interactions between an agent and its environment, composed of interactions that take place through sensors and actuators. Both sensors and actuators are means for exploiting affordances in the environment; in addition, they carry specific semantic information for the achievement of agent's goals. Evolution seizes affordances by means of heritable variations and selection, and therefore, what is relevant for the organism provides an evolutionary advantage and is selected. Conversely, in artificial systems, the designers usually decide in advance what is relevant to the system and consequently equip it with proper devices. This difference between natural and artificial systems has two main consequences: the first one is that the interactions taking place between an organism and the environment are inherently *open*, as they depend on the emergence of meaning from affordances; the second is that the design of systems able to effectively attain autonomous behavior in a given environment should include the possibility of autonomous development of sensors and actuators. In the following sections, we will discuss these two aspects.

4 On the Openness of Organism–Environment Interactions

The interactions between an organism and its environment that arise from the use of sensing and acting organs are a function of the evolution of those specific organs. The emergence of these organs through evolution is enabled by affordances, which are seized by heritable variations and selection. A cardinal property of affordances is that they are inherently unpredictable (Kauffman, 2019; Kauffman & Roli, 2021) and so is the emergence of sensing and acting organs. Therefore, the description of this kind of interactions is essentially incomplete. New interactions may arise in time, either by the emergence of new organs or new ways for using the current ones. For example, flight feathers initially evolved as thermal insulators, but were then co-opted for the new function of flight (Prum & Brush, 2002). The evolutionary adaptation to new environmental conditions can also be attained by *phenotypic plasticity*, whereby adaptation occurs without requiring a genetic variation. This phenotypic variant can be inherited if produced by epigenetic differences in gene expressions (Thorson et al., 2017), or it can bias the subsequent selection of favorable genetic traits (West-Eberhard, 2005).

This incompleteness is the very source of actual open-ended evolution, which is sustained by the openness of interactions between organisms and the environment. This huge space of possible relations becomes even wider if we consider the interactions among organisms (Roli & Kauffman, 2022), where interactions appear in a profusion of manifestations, from mutualistic to parasitic relations. A prominent example of this creative incompleteness of interactions is provided by symbiosis (Margulis, 2008), which can be interpreted as the result of combinations of seized affordances, in which species interaction provides reciprocal and long-term coupling sustaining their life. New super-organisms, *holobionts*, may form as a consequence of dynamically evolving symbiotic interactions between different species.

5 Toward Epistemic Autonomous Machines

In mid 1950s, Gordon Pask, one of the greatest actors in early cybernetics, made several experiments with electrochemical devices with the aim of building physical systems able to develop their own sensors, hence their own relevance criteria for perceiving the world (Pask, 1958a, 1958b). Despite the outstanding intuition that autonomous systems should be able to develop their own devices for interacting with the physical world, those pioneering explorations got somehow lost in oblivion and were not further developed. However, the significance of Pask's work has been effectively discussed by Cariani (1993), who advocates a cybernetic-semiotic frame for the study and design of artificial systems. The core statement is that "sensors and effectors are the crucial points at which a real-world situation is encoded into a symbolic representation and at which an action-decision is transformed from symbolic representation into physical action. It is only by virtue of actual connection to the world via sensors and actuators that symbolic representations become semantically grounded" (Cariani, 1998). The inherent openness of interactions between agents and the physical world should then be considered as the space of possibilities for the development of epistemic emergence. In this view, actual autonomous systems should be able be attain epistemic autonomy, i.e., the capability of tuning their interactions with the world in such a way that relevant pieces of information are extracted and relevant operations are performed.

There are of course major technological issues for achieving such a kind of artificial systems. However, it is possible to reduce the generality of the scenario and address these goals with devices that can be more easily built and work in safe conditions. A viable way has been recently proposed, which consists in equipping robots with a network controller, e.g., a Boolean network (Kauffman, 1993), and let the robot adapt the connections from sensors to the network and from the network to the actuators according to an intrinsic reward function (Braccini et al., 2022, 2023). In Fig. 1, a schematic representation of the controller and its connections to the physical world is depicted. In this way, the interactions with the environment are adapted so as to exploit the relevant features of the intrinsic dynamics of the system. Sensors and actuators are provided in advance; therefore, there is no actual evolution of sensing and acting organs, but it is their relation with the controller that is left open to adapt. Recent results along this line of research have shown that robots controlled by Boolean networks can adapt their interactions with sensors and actuators so as to be able to perform simple tasks such as phototaxis, collision avoidance, and foraging (Braccini et al., 2022), without requiring to adapt the controller.

A further step in this direction consists in the artificial evolution of robotic holobionts, i.e., multi-robot systems composed of heterogeneous robots, each with specific sensing and acting abilities. The key mechanisms that enable super-organisms to achieve high level of complexity behaviors are to be found in the phenotypic heterogeneity of the individuals—observable, e.g., in the form of different morphologies and behavioral specialization (O'Shea-Wheller et al., 2021)—and in the possibility of opportunistically exploiting the behaviors and the information processing capabilities of the peers, in a model of interaction that resembles that of a dynamical network of possibly different computational nodes that exchange information to achieve shared goals. An illustration of the elements and the interactions occurring among them is presented in Fig. 2. This approach makes it possible to evolve composite artificial systems that autonomously choose and tune their interactions with the environment, by combining the different capabilities of the individual

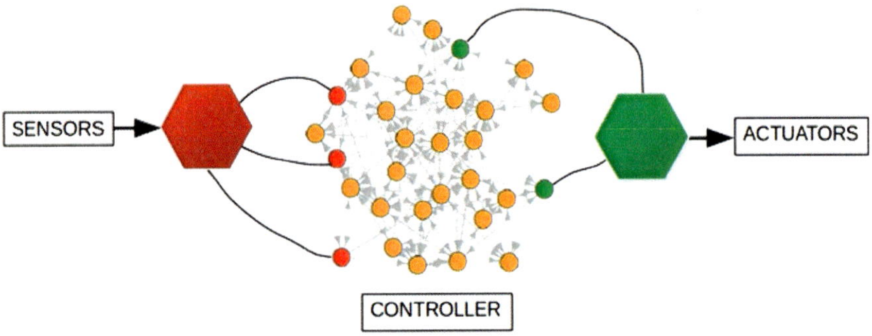

Fig. 1 Schematic representation of the Boolean network controller and its couplings with robot sensor and actuators. While the network is kept unchanged, the connections from sensors to the network nodes and from network nodes to the actuators are subject to modification during the "life" of the robot. These changes make it possible to adapt the interaction between the robot and the environment so as to maximize a given utility function

robots. Also in this case, the initial building blocks are defined in advance, but their combinations are open and depend upon the tasks that the final composite robotic super-organism has to perform.

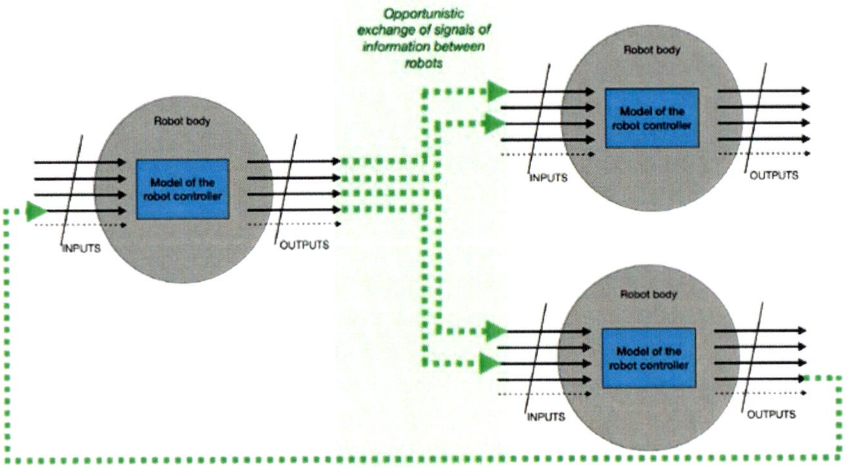

Fig. 2 Schema of the interactions occurring between the different robots composing the holobiont. While controllers are not supposed to change, their interactions, as well as the actual combination of different robots, is subject to an adaptive process

6 Conclusion and Future Work

In this contribution, we have reviewed the open nature of interactions between agents—organisms and robots—and their environment from a cybernetic perspective. This viewpoint makes it possible to emphasize the role played by sensors and actuators in actualizing these interactions, and the crucial role of affordances, both in natural and artificial scenarios. The impossibility of predicting affordances is at the core of the inherent incompleteness of any description of this kind of interaction. However, it is also the source of novelty that can arise when interactions can be developed autonomously. We believe that the message from the early cybernetic scholars is still of foremost importance for studying and designing organisms and artificial systems from a systemic perspective. The design of robots able to adapt their own interactions with the physical world, with the perspective of building robotic super-organisms, builds upon these considerations, which should also be taken as advises and warnings in the deployment of such artificial systems. The wide space of possibilities that originates from open interactions has a downside to be faced: the impossibility of fully predicting the behavior of such systems and the outcome of their interactions with human beings. A perspective acknowledging the incompleteness of our knowledge about the actual dynamics of such devices may provide an honest approach to the understanding and the design of complex artificial systems.

References

Braccini, M., Roli, A., Barbieri, E., & Kauffman, S. (2022). On the criticality of adaptive boolean network robots. *Entropy, 24*(10), 1368:1–1368:21

Braccini, M., Roli, A., & Kauffman, S. (2023). A novel online adaptation mechanism in artificial systems provides phenotypic plasticity. *Artificial life and evolutionary computation: 15th Italian workshop, WIVACE 2021, Winterthur, Switzerland, September 15–17, 2021* (pp. 121–132). Springer.

Braitenberg, V. (1986). *Vehicles: Experiments in synthetic psychology*. MIT Press.

Brambilla, M., Ferrante, E., Birattari, M., & Dorigo, M. (2013). Swarm robotics: A review from the swarm engineering perspective. *Swarm Intelligence, 7*, 1–41.

Brooks, R. (1990). Elephants don't play chess. *Robotics and Autonomous Systems, 6*(1–2), 3–15.

Campbell, C., Olteanu, A., & Kull, K. (2019). Learning and knowing as semiosis: Extending the conceptual apparatus of semiotics. *Sign Systems Studies, 47*(3–4), 352–381.

Cariani, P. (1993). To evolve an ear. Epistemological implications of Gordon Pask's electrochemical devices. *Systems Research, 10*(3), 19–33.

Cariani, P. (1998). Towards an evolutionary semiotics: The emergence of new sign-functions in organisms and devices. In *Evolutionary systems* (pp 359–376). Springer.

Gibson, J. (1966). *The senses considered as perceptual systems*. Houghton Mifflin.

Heras-Escribano, M. (2019). *The philosophy of affordances*. Springer.

Hoffmeyer, J. (2008). *Biosemiotics: An examination into the signs of life and the life of signs*. University of Chicago Press.

Jamone, L., Ugur, E., Cangelosi, A., Fadiga, L., Bernardino, A., Piater, J., & Santos-Victor, J. (2016). Affordances in psychology, neuroscience, and robotics: A survey. *IEEE Transactions on Cognitive and Developmental Systems, 10*(1), 4–25.

Kauffman, S. (1993). *The origins of order: Self-organization and selection in evolution*. Oxford University Press.
Kauffman, S. (2019). *A world beyond physics: The emergence and evolution of life*. Oxford University Press.
Kauffman, S., & Roli, A. (2021). The world is not a theorem. *Entropy, 23*(11), 1467.
Kull, K. (1999). Biosemiotics in the twentieth century: A view from biology. *Semiotica, 127*(1–4), 385–414.
Margulis, L. (2008). *Symbiotic planet: A new look at evolution*. Basic books.
Metta, G., & Fitzpatrick, P. (2003). Early integration of vision and manipulation. *Adaptive Behavior, 11*(2), 109–128.
Minati, G. (2022). Interactions and interaction mechanisms as generators of complex systems. *WSEAS Transactions on Proof, 2*, 96–107.
Nehmzow, U. (2008). *Robot behaviour: Design, description, analysis and modelling*. Springer.
O'Shea-Wheller, T., Hunt, E., & Sasaki, T. (2021). Functional heterogeneity in superorganisms: Emerging trends and concepts. *Annals of the Entomological Society of America, 114*(5), 562–574.
Pask, G. (1958a). The growth process inside the cybernetic machine. In *Second International Conference on Cybernetics* (pp. 765–794).
Pask, G. (1958b). Physical analogues to the growth of a concept. *Mechanization of Thought Processes, Symposium, 10*, 765–794.
Pfeifer, R., & Scheier, C. (1999). *Understanding intelligence*. MIT Press.
Prum, R., & Brush, A. (2002). The evolutionary origin and diversification of feathers. *The Quarterly Review of Biology, 77*(3), 261–295.
Roli, A., & Braccini, M. (2018). Attractor landscape: A bridge between robotics and synthetic biology. *Complex Systems, 27*, 229–248.
Roli, A., & Kauffman, S. (2020). Emergence of organisms. *Entropy, 22*(10), 1163, 1–12.
Roli, A., & Kauffman, S. (2022). The hiatus between organism and machine evolution: Contrasting mixed microbial communities with robots. *Bio Systems, 222*, 104775.
Roli, A., Jaeger, J., & Kauffman, S. (2022). How organisms come to know the world: Fundamental limits on artificial general intelligence. *Frontiers in Ecology and Evolution, 9*, 1035.
Simon, H. (2019). *The sciences of the artificial*. MIT Press.
Thorson, J., Smithson, M., Beck, D., Sadler-Riggleman, I., Nilsson, E., Dybdahl, M., & Skinner, M. (2017). Epigenetics and adaptive phenotypic variation between habitats in an asexual snail. *Scientific Reports, 7*(1), 1–11.
von Uexküll, J. (2010). *A foray into the worlds of animals and humans: With a theory of meaning* (J. D. O'Neil, Trad.). University of Minnesota Press (originally published in German 1934).
Von Foerster, H. (2003). Cybernetics of epistemology (Chap. 9). In *Understanding: Essays on cybernetics and cognition* (pp. 229–246). Springer.
Walsh, D. (2015). *Organisms, agency, and evolution*. Cambridge University Press.
West-Eberhard, M. (2005). Developmental plasticity and the origin of species differences. *Proceedings of the National Academy of Sciences, 102*(suppl. 1), 6543–6549.
Wiener, N. (1948). *Cybernetics: Or control and communication in the animal and the machine*. MIT Press.

Complexity in Human Systems

The Association Between Stress and Well-Being with Resilience and Coping in University Students During Covid-19 Pandemic: A Longitudinal Network Analysis

Roberta Renati, Natale Salvatore Bonfiglio, and Dolores Rollo

Abstract Coping and resilience, which are the ways in which we cope with stress and difficult times, represent two separate but interconnected constructs. The literature has not yet fully clarified the extent to which the two concepts overlap or differ. The present study aims to investigate the relationship between coping and resilience and their interconnections with well-being and stress in the Covid-19 pandemic, through a network analysis. Participants aged 18–24 are university students who completed an online self-report battery measuring resilience, coping, stress, and well-being at two different times T0 (during lockdown from Covid-19) and T1 (one year after the first completion). The Resilience Scale for Adults (RSA), the Brief COPE, the Perception of Stress Scale (PSS), and the Warwick–Edinburgh Mental Well-being Scale (WEMWS) were administered. Through a network analysis, it was found that coping and resilience show associations with measures of well-being, but less with stress. Specifically, the RSA subscales "self-perception" and "perception of the future", and adaptive coping strategies show major association and direct relationships well-being. The results, therefore, show that coping and resilience are distinct but clearly related constructs, and that resilience presents the most weight influence. Comparing the connectedness networks at T0 and T1 shows that both constructs are less important in reducing stress levels but more important in improve well-being. The results obtained in this study may have important implications for structuring specific interventions in periods of difficulty and stress for the university student population.

R. Renati
Department of Pedagogy, Psychology, Philosophy, University of Cagliari, Cagliari, Italy
e-mail: roberta.renati@unica.it

N. S. Bonfiglio (✉)
Statistical Service, IRCCS, Centro San Giovanni di Dio Fatebenefratelli of Brescia, Brescia, Italy
e-mail: nbonfiglio@fatebenefratelli.it

D. Rollo
Department of Medicine and Surgery, University of Parma, Parma, Italy
e-mail: dolores.rollo@unipr.it

G. Minati and M. Pietronilla Penna (eds.), *Multiple Systems*, Contributions to Management Science, https://doi.org/10.1007/978-3-031-44685-6_16

1 Introduction

Several research have shown how the Covid-19 pandemic produced serious consequences on social and psychological well-being (Dey & Loewenstein, 2020). The spread of Covid-19, compared to similar events experienced to date, was characterized by an almost immediate globality of the phenomenon that made it unprecedented and found modern humans completely unprepared and fragile, at the mercy of events.

Serafini et al. (2020), in a recent review, pointed out that among the first psychological responses to such an event are increased levels of anxiety, stress, altered mood and levels of irritability. But frustration and boredom also come into play, which along with social isolation due to lockdowns, led people to experience a deep sense of loneliness.

The consequences also affected all segments of the population indiscriminately, including adolescents and young adults. More than a third of them, in fact, reported high levels of loneliness, and nearly half of 18–24-year-olds felt lonely during lockdown, also experiencing depressive symptoms (Loades et al., 2020). Among the youth, the student category suffered from severe psychological distress, with high levels of anxiety in students due to lack of pleasure and satisfaction in taking online courses and the absence of interaction with peers (Rohman et al., 2020).

The lack of contact with one's peers and social connections has had a significant impact on the mental health of this segment of the population (Berkman, 1983). In fact, a healthy development of a social network allows the individual to have access to what is called social support, such as instrumental support (e.g., practical help with tasks and activities), informational support (e.g., advice and opinions), and emotional support (e.g., feeling loved and listened to). It also acts in complicity with other psychosocial mechanisms, such as making a good network of friends and family members, which are useful in supporting general well-being and reducing physiological arousal and consequently stress perception (Umberson et al., 2010).

Among the many risks factors consequent to the lockdown period that have affected, this very population group is specifically the social immobility resulting from the lockdowns and the economic decline that will fall on their future. Uncertainty, isolation, and fear for one's own and others' health induced the elevation of stress levels and the exacerbation of prior mental issues, including internalizing symptoms and anger (Reger et al., 2020).

Indeed, one must consider that adolescents and young adults have faced this difficult historical period at an already challenging time for any individual, being faced with many existential turning points (school, work, and relationships) (Arnett, 2000). It is exactly the latter that become fundamental to the adolescent's identity process, which is based and founded on affective investments and social relationships outside the family unit (Sica et al., 2018). In this context, blocking and restraining measures can become a risk factor and a form of regression that triggers even more negative psychological health outcomes.

In fact, a study conducted on subjects between the ages of 18 and 30 during the first four weeks of lockdown showed an increase in internalizing symptoms (anxiety, isolation, and depression) and externalizing symptoms (aggressive and rule-breaking behaviors) (Parola et al., 2020). These data find correlation with what has already been observed in studies prior to this crisis period that point out how the inability to have social contacts and behaviors and the physical distance from one's social network intensify negative feelings of distress and strong anxiety (Boffo et al., 2012).

A strong influence on personal well-being under conditions of distress is played by the coping mechanisms that each individual is able to implement, i.e., the adaptive strategies and ways that the individual is able to engage in dealing with emotional and interpersonal problems with the aim of mitigating psychological stress.

Several studies have, therefore, analyzed such coping mechanisms during the pandemic period and the strong correlation between coping and psychological well-being, pointing out that in individuals who implemented negative coping styles, higher levels of distress were then reported (Wang et al., 2020).

In a situation such as the blockade caused by the pandemic, in general, every individual experienced feeling of anxiety and fear. In the youth population, with previous vulnerabilities and unfavorable social conditions, these risks on mental health were even greater given the condition of emotional immaturity (Cruz et al., 2020).

Chew et al. (2020) in a recent review, summarized the coping mechanisms and psychological and adaptive responses of the population during previous epidemics and climate disasters. Problem-solving skills, seeking social support, and maintaining positive esteem toward, for example, one's own country's government and health care system, were found to be the most useful strategies for coping best during times of crisis. In contrast, an approach marked by distraction-seeking, denial, and avoidance of the situation correlated negatively with later reported stress levels.

Compared to earlier social crises, moreover, the Covid period was characterized by the need for the individual's isolation, resulting in an unprecedented effort to maintain resilience (Polizzi et al., 2020).

An Italian study focused on adolescents by investigating their ability to positively or negatively re-read the new context in which they found themselves living. The results revealed that adolescents were more likely to report their experience in a negative light (Fioretti & Smorti, 2015). The negative aspects highlighted focused on the limitation of autonomy and the difficulty in expressing and discovering their new identity. However, in many cases, many young people were also able to reinterpret their condition in a positive light, showing evidence of being able to make use of coping resources. These re-interpretations cantered on the possibility of rediscovering themselves through new moments favoring introspection, on the one hand, and re-evaluating family relationships by finding themselves sharing the same spaces for much longer periods of time, on the other. These aspects were seen as part of a broad process of personal growth and helped the maintenance of the children's well-being (Fioretti et al., 2020).

A further study that investigated what may be protective factors for overcoming stressful events and experiencing less psychological pressure pointed out that, in fact, older age correlates with better awareness, cognitive resilience, and emotional balance that allow for more optimal coping strategies (Conversano et al., 2020).

These data highlight how important it is to pay attention to the mental health of younger people as although they have the resources to cope, prevention, support, and care interventions are still needed.

2 Methodology

Three hundred and thirty-nine subjects completed an online survey via a Web link. The surveys were sent in April 2020 (T0) and May 2021 (T1). All subjects who responded at T0 and 64 at T1 were included in the analysis. All subjects were aged between 18 and 24 years, of which 36 were male.

2.1 Instruments

The Brief COPE (BC) is a shortened version of the Coping Orientation to Problem Experienced, also called COPE Inventory, a questionnaire for investigating coping responses to stressful or crisis situations. The BC consists of 14 scales of 2 items each, and each scale refers to two conceptually different coping mechanisms, a coping characterized by adaptive strategies (BC_AD), such as seeking emotional support, and one characterized by non-adaptive strategies (BC_NAD), such as turning to substance use. Using a Likert scale from 1 to 4, the subject can indicate how habitually he or she does or does not engage in these behaviors. The internal validity of the Brief COPE proved to be similar to the full version (COPE) and with regard to reliability the Cronbach's alpha values for the items all exceeded the minimum acceptable threshold of 0.50.

The Warwick–Edinburgh Mental Well-Being Scale (WEMWBS) is a scale consisting of 12 items, all positively worded and relating to aspects of mental well-being as a broad concept including emotional, cognitive, and evaluative aspects. Each item is answered on a scale of 1–5 steps to define the frequency with which these aspects were felt and experienced, and is asked to focus on the specific period of two weeks prior to the administration of the questionnaire. The WEMWBS showed good content validity and an alpha value of 0.89 was reported for the student sample and 0.91 for the general population sample.

The Resilience Scale for Adults (RSA) is a multidimensional scale consisting of 33 items designed to measure six different dimensions of resilience: self-perception (RSA_PS) and perception of the future (RSA_PF), one's structured style (RSA_SS), social resources and skills (RSA_SR), and family cohesion (RSA_FC). The subject is asked to indicate by means of a five-step semantic differential the statement that he/she feels is most related to his/her habitual behavior. The validation indices for each subscale range from 0.67 to 0.90, with test–retest values from 0.69 to 0.84.

The Perceived Stress Scale (PSS), is a scale to measure the perception of stress. It consists of ten items that investigate how often the subject in the last two weeks has perceived his or her life as unpredictable, uncontrollable, and burdensome, and consequently provide an overall picture of the possible levels of stress arising. The alpha reliability coefficient was 0.86 in the three different subject groups. The test–retest correlation was 0.85 in the samples retested after two days and only 0.64 for the subjects retested after six weeks.

2.2 *Data Analysis*

To investigate the relationship between coping as measured at T0 by the BC, resilience as measured at T0 by the RSA, stress at T1 as measured by the PSS, and well-being at T1 as measured by the WEMWBS, we estimated a combined partial correlation network including ten nodes. All edges were interpreted as polychoric partial correlations (− 1 to 1). The network was estimated using Gaussian–Markov random field estimation via LASSO graph, where the optimal regularization parameter was selected using the extended Bayesian information criterion (EBIC).

To study the centrality of nodes within the network, i.e., the role of each node within the network, three common measures of centrality were estimated using the Jasp package (Love et al., 2019): (1) node strength, which is the direct connection of a node to the network, calculated as the sum of the absolute weights of all edges of a given node with all others, (2) closeness, which is the indirect connection of a node to the network, calculated as the sum of the inverse of all shortest path lengths between a node and all others, and (3) betweenness, which is the indirect connection of a node to the network, calculated as the number of times a node is on the shortest path connecting two other nodes in a network. In general, more central nodes have higher centrality values.

To examine the accuracy of the network, the accuracy of the edge weights was assessed by bootstrapping (1000 iterations) the 95% confidence intervals (CIs) around the edge weights. Smaller CIs indicate higher accuracy.

3 Results

To investigate the relationship between the four constructs considered, a correlation network was estimated. Figure 1 shows a visualization of the correlation network. Overall, the network shows clear associations both within and between scales. Zooming in on the nodes, strong positive associations emerged between post_WEB and RSA_PF, BC_AD, and RSA_PS. Negative association emerged between post_WEB with both RSA_SS and BC_NAD. Regarding stress subscale, negative association emerged with both RSA_FC and RSA_SR, and with post_WEB. Positive association emerged with RSA_SS (Table 1).

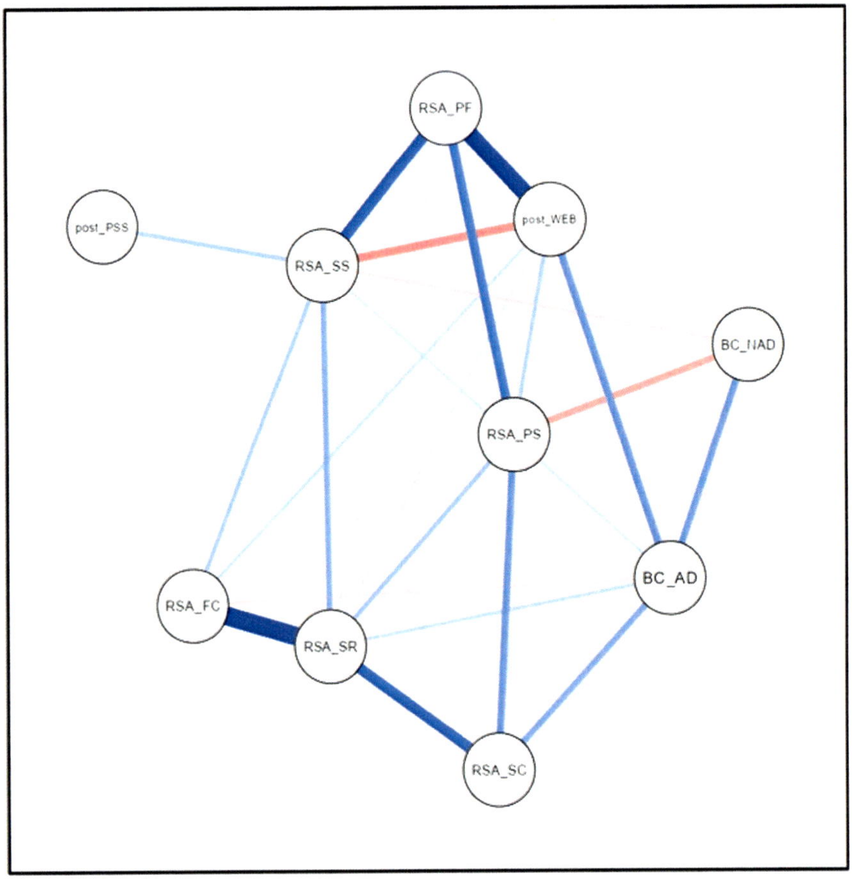

Fig. 1 Network associations between subscales. *Legend* BC_AD = adaptive coping strategies; BC_NAD = non-adaptive coping strategies; RSA_PS = perception of self-subscale of RSA; RSA_PF = perception of future subscale of RSA; RSA_SS = structured style scale of RSA; RSA_SC = social competencies subscale of RSA; RSA_FC = family cohesion subscale of RSA; RSA_SR = social resources subscale of RSA; post_PSS = perception of stress subscale; post_WEB = well-being subscale

It is interesting to note the presence of stronger correlations for the node perception of the future and structured style, and that the relationship between structured style and well-being is negative.

To analyze the centrality of nodes within the network, the strength, closeness, and betweenness of nodes were estimated (Fig. 2).

Regarding node strength, the nodes with the highest strength are RSA_SR and post_WEB, while the one with the lowest strength is post_PSS. These nodes with high strength are probable to interact with other nodes in the network because they (1) predict, (2) are predicted, or (3) predict and are predicted by other nodes in the network (Bringmann et al., 2019).

Table 1 Weights matrix

Variable	Network									
	BC_AD	BC_NAD	RSA_PS	RSA_PF	RSA_SS	RSA_SC	RSA_FC	RSA_SR	post_PSS	post_WEB
BC_AD	0.000	0.173	0.042	0.000	0.071	0.153	− 0.071	0.087	0.000	0.185
BC_NAD	0.173	0.000	− 0.161	0.000	− 0.084	0.000	0.000	− 0.060	0.051	− 0.047
RSA_PS	0.042	− 0.161	0.000	0.226	0.000	0.175	0.000	0.111	0.000	0.098
RSA_PF	0.000	0.000	0.226	0.000	0.271	0.000	0.021	0.053	0.000	0.401
RSA_SS	0.071	− 0.084	0.000	0.271	0.000	0.000	0.102	0.141	0.107	− 0.197
RSA_SC	0.153	0.000	0.175	0.000	0.000	0.000	0.008	0.260	0.000	0.052
RSA_FC	− 0.071	0.000	0.000	0.021	0.102	0.008	0.000	0.444	− 0.022	0.081
RSA_SR	0.087	− 0.060	0.111	0.053	0.141	0.260	0.444	0.000	− 0.017	0.060
post_PSS	0.000	0.051	0.000	0.000	0.107	0.000	− 0.022	− 0.017	0.000	− 0.024
post_WEB	0.185	− 0.047	0.098	0.401	− 0.197	0.052	0.081	0.060	− 0.024	0.000

Legend BC_AD = adaptive coping strategies; BC_NAD = non-adaptive coping strategies; RSA_PS = perception of self-subscale of RSA; RSA_PF = perception of future subscale of RSA; RSA_SS = structured style scale of RSA; RSA_SC = social competencies subscale of RSA; RSA_FC = family cohesion subscale of RSA; RSA_SR = social resources subscale of RSA; post_PSS = perception of stress subscale; post_WEB = well-being subscale

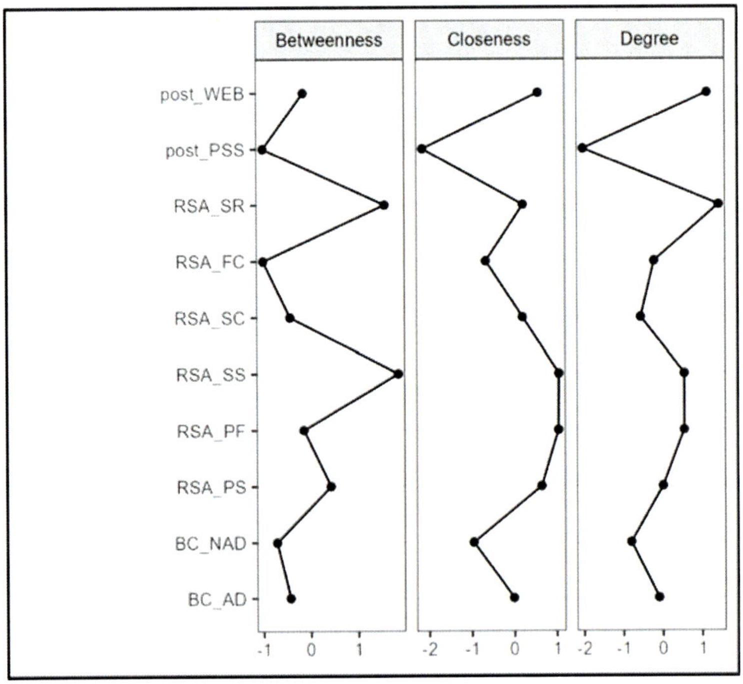

Fig. 2 Centrality plot

Regarding the closeness of nodes, those with the greatest closeness are RSA_SS, RSA_PS, and RSA_PF, while the one with the least closeness is post_PSS. Regarding the betweenness of the nodes, the nodes with the highest betweenness were RSA_SS and RSA_SR, and the one with the lowest betweenness was post_PSS. Interestingly, stress represents the node with the least proximity and the least strength on the others. While especially perception of future, social resources and structured style represent dimensions of resilience with greater closeness and strength than the other nodes and especially with well-being (Table 2).

To verify the ***accuracy*** of the estimated network, a bootstrap procedure was performed on the 95% CIs around the edge weights (Fig. 3). The results show reasonably small bootstrap ICs around most of the estimated edge weights, indicative of high accuracy. Larger bootstrap ICs imply that the interpretation of edge order in the network must be done carefully.

Table 2 Centrality measures per variable

	Network			
Variable	Betweenness	Closeness	Strength	Expected influence
BC_AD	− 0.428	− 0.010	− 0.099	0.284
BC_NAD	− 0.713	− 0.955	− 0.801	− 1.849
RSA_PS	0.428	0.652	0.009	− 0.129
RSA_PF	− 0.143	1.047	0.546	1.205
RSA_SS	1.855	1.057	0.550	− 0.355
RSA_SC	− 0.428	0.206	− 0.555	0.307
RSA_FC	− 0.999	− 0.660	− 0.212	0.069
RSA_SR	1.570	0.218	1.433	1.500
post_PSS	− 0.999	− 2.140	− 2.011	− 1.231
post_WEB	− 0.143	0.585	1.140	0.199

4 Discussion

Based on the findings of the network analysis and the indexes of centrality, strength, and betweenness, and shifting the focus to well-being and stress as outcomes at T1, Self-perception, perception of the future, and structured style appear to be the most significant nodes in relation to well-being. The use of coping strategies, moreover, seems to have a role in maintaining a condition of well-being but not in reducing stress, the association of which is "far" from the point of view of distance between the nodes. One element to consider to get a more complete picture with respect to young adults' experience, perception, and well-being seems to be that of future expectations. This result seems in line with findings in the literature. For example, a Polish study conducted in the early stages of the pandemic found that in subjects who stated that they generally had high baseline expectations and considered themselves satisfied with their lives and attributed meaning to them, the level of anxiety related to the occurrence of Covid was much lower (Trzebiński et al., 2020). Analysis suggests that this is because such positive global feelings act as protection in that they predispose to having encouraging expectations about the world and one's future. The study, however, shows that an unpredictable social environment and scarcity of resources, two conditions easily identified at the time of the pandemic crisis, become

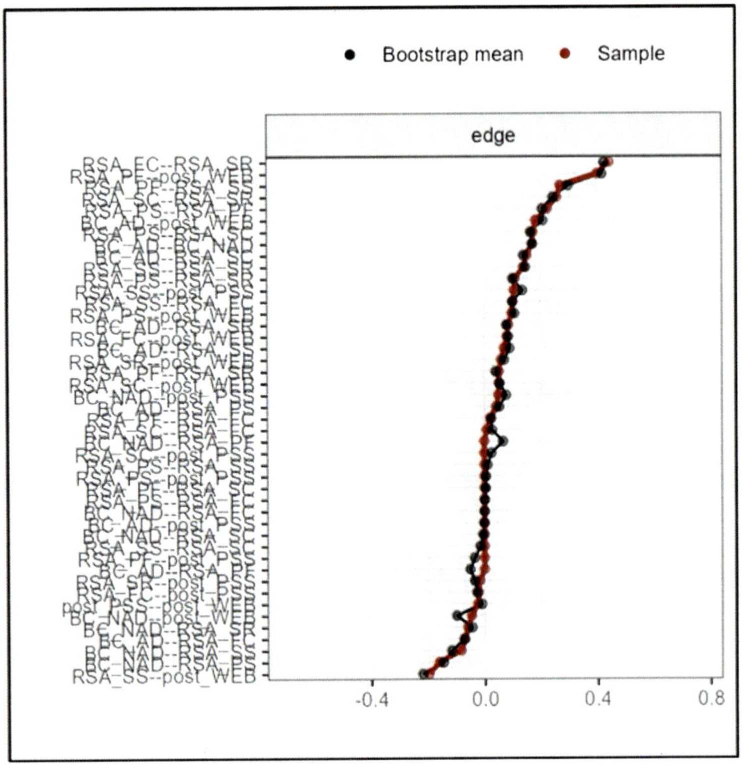

Fig. 3 Accuracy graph

predictors of negative cascading strategies, such as a short-term view of life and lack of investment in the future. Therefore, it is important to investigate the types of social experiences in an era of viral panic, as they can consolidate these global feelings on the negative side and influence the individual's life. In our research, however, structured style and social resources were found to be key nodes in conveying well-being and the use of adaptive strategies. It is precisely this last result that shows that it becomes necessary to support especially the youth population in dealing with the uncertainties associated with the post-Covid period. Adolescents and young adults experienced the quarantine, compared to the adult segment of the population, associating it with a strong sense of insecurity and vagueness toward their near future specifically because they were not already adults, and with a conscious individuality and a structured lifestyle, the changes faced in order to undergo the imposed restrictions had a greater impact (Commodari & La Rosa, 2020).

This result is also interesting if we consider that the differences in perception of pandemic based on registry age is different between young and adult people. The latter have a greater ability to think realistically about the evolution of the situation, but also in assessing and predicting their own future emotions, as the greater life experience is an important coping mechanism specific to adulthood (Ceccato et al., 2021). The subjects in our sample, however, seemed to have resiliently used their perception of the future (and thus hope for the future) and social resources and skills. In general, adolescents and youth tend to be more optimistic toward the future and expect positive outcomes, compared to adults. This factor allows them to better regulate their emotions in times of crisis. In contrast, less mature individuals are significantly more shaken by seeing their options for action that could affect the future limited; this may lead them to have an unclear view of the future and amplify the perception of negative emotions (Ceccato et al., 2021).

The literature considers generalized fear and pervasive anxiety to be two among the earliest responses recorded to be associated with previous infectious disease outbreaks (Maunder et al., 2003). These responses may occur more easily and more severely in the presence of certain risk factors such as poor basic resources (water, food, and clothing) and inadequate information about the events. In contrast, protective factors that can reduce the extent of possible emotional disturbances include: active social support and preventive strategies such as effective communication. This was confirmed by the results of our study in which precisely structured style and social skills seem to have represented two important nodes in relation to the use of adaptive coping strategies. This confirms how adequate psychological listening and helping services are. The implementation of telemedicine, especially with more fragile patients and subjects with psychological problems, may also be the best way to deal with situations such as those typical of a pandemic and to stimulate in subjects the use of adaptive coping strategies (Serafini et al., 2020).

5 Conclusion

Considering the peculiar challenges of adolescence and early adulthood, one of the most crucial acquisitions is precisely that of social skills, which, together with the individual's attitudes and skills, helps ensure a successful transition to adult life. Indeed, young people are required to find their own identity and discover a new social role as they make decisions about their education and future work.

The first substantial relationship that undergoes change is the one with parents. This specific relationship is highly correlated with the well-being of the young person because in addition to material support, the family should provide emotional support and perform a modeling function. On the other hand, in situations where all this is lacking, one's relationship with the family becomes a risk factor.

The results of our study showed that family cohesion represented the least central and most distant node with respect to well-being and the use of adaptive coping strategies. Thus, personal, and social resources seem to carry the greatest weight in activating positive outcomes, while family seems to be relegated to a secondary role. It is the friendship and peer group relationships themselves that make the difference. These findings are confirmed by several studies regarding the danger of an absence of social connection (Zarrett & Eccles, 2006). Loneliness increases an individual's sensitivity and impairs executive functioning, sleep quality, and well-being in toto (mental and physical). In addition to this, a person who feels lonely will not only experience sadness, but also a strong sense of general insecurity, which in turn will activate survival mechanisms that make them stay alert as if every external stimulus can be identified as a danger. This mechanism becomes counterproductive when the individual, although feeling lonely, when able to relate to others, perceives only the negative aspects of the interaction (Cacioppo, 2014).

The interesting results obtained in this study should be read with caution because of some limitations. First, the sample size is unbalanced between T0 and T1 and very low at T1 compared with T0. In addition, there is a high prevalence of female subjects. Future research under crisis conditions could replicate the same results on a more balanced sample. In addition, due to the low adherence of subjects at T1, a second follow-up (T2) could not be planned. Finally, subject selection had several biases, such as involving only students who were directly attending the online courses of the authors of this study.

References

Arnett, J. J. (2000). Emerging adulthood: A theory of development from the late teens through the twenties. *American Psychologist, 55*(5), 469–480.

Berkman, L. F. (1983). Health and way of living. The Alameda County Study.

Boffo, M., Mannarini, S., & Munari, C. (2012). Exploratory structure equation modeling of the UCLA loneliness scale: A contribution to the Italian adaptation. *TPM: Testing, Psychometrics, Methodology in Applied Psychology, 19*(4).

Bringmann, L. F., Elmer, T., Epskamp, S., Krause, R. W., Schoch, D., Wichers, M., Wigman, J. T., & Snippe, E. (2019). What do centrality measures measure in psychological networks? *Journal of Abnormal Psychology, 128*(8), 892.

Cacioppo, J. T., & Cacioppo, S. (2014). Social relationships and health: The toxic effects of perceived social isolation. *Social and Personality Psychology Compass, 8*(2), 58–72.

Ceccato, I., Palumbo, R., Di Crosta, A., La Malva, P., Marchetti, D., Maiella, R., Verrocchio, M. C., Marin, A., Mammarella, N., Palumbo, R., & Di Domenico, A. (2021). Age-related differences in the perception of COVID-19 emergency during the Italian outbreak. *Aging & Mental Health, 25*(7), 1305–1313.

Chew, Q. H., Wei, K. C., Vasoo, S., Chua, H. C., & Sim, K. (2020). Narrative synthesis of psychological and coping responses towards emerging infectious disease outbreaks in the general population: Practical considerations for the COVID-19 pandemic. *Singapore Medical Journal, 61*(7), 350.

Commodari, E., & La Rosa, V. L. (2020). Adolescents in quarantine during COVID-19 pandemic in Italy: Perceived health risk, beliefs, psychological experiences and expectations for the future. *Frontiers in Psychology, 11*, 559951.

Conversano, C., Di Giuseppe, M., Miccoli, M., Ciacchini, R., Gemignani, A., & Orrù, G. (2020). Mindfulness, age and gender as protective factors against psychological distress during Covid-19 pandemic. *Frontiers in Psychology, 11*, 1900.

Dey, M., & Loewenstein, M. A. (2020). How many workers are employed in sectors directly affected by COVID-19 shutdowns, where do they work, and how much do they earn? *Monthly Labor Review*, 1–19.

Fernández Cruz, M., Álvarez Rodríguez, J., Ávalos Ruiz, I., Cuevas López, M., de Barros Camargo, C., Díaz Rosas, F., González Castellón, E., González González, D., Hernández Fernández, A.,

Ibáñez Cubillas, P., & Lizarte Simón, E. J. (2020). Evaluation of the emotional and cognitive regulation of young people in a lockdown situation due to the Covid-19 pandemic. *Frontiers in Psychology, 11.*

Fioretti, C., & Smorti, A. (2015). How emotional content of memories changes in narrative. *Narrative Inquiry, 25*, 37–56.

Fioretti, C., Palladino, B. E., Nocentini, A., & Menesini, E. (2020). Positive and negative experiences of living in COVID-19 pandemic: Analysis of Italian adolescents' narratives. *Frontiers in Psychology, 11*, 3011.

Loades, M. E., Chatburn, E., Higson-Sweeney, N., Reynolds, S., Shafran, R., Brigden, A., Linney, C., McManus, M. N., Borwick, C., & Crawley, E. (2020). Rapid systematic review: The impact of social isolation and loneliness on the mental health of children and adolescents in the context of COVID-19. *Journal of the American Academy of Child & Adolescent Psychiatry, 59*(11), 1218–1239.

Love, J., Selker, R., Marsman, M., Jamil, T., Dropmann, D., Verhagen, J., Ly, A., Gronau, Q. F., Šmíra, M., Epskamp, S., Matzke, D., & Wagenmakers, E. J. (2019). JASP: Graphical statistical software for common statistical designs. *Journal of Statistical Software, 88*, 1–17.

Maunder, R., Hunter, J., Vincent, L., Bennett, J., Peladeau, N., Leszcz, M., Sadavoy, J., Verhaeghe, L. M., Steinberg, R., & Mazzulli, T. (2003). The immediate psychological and occupational impact of the 2003 SARS outbreak in a teaching hospital. *CMAJ, 168*(10), 1245–1251.

Parola, A., Rossi, A., Tessitore, F., Troisi, G., & Mannarini, S. (2020). Mental health through the COVID-19 quarantine: A growth curve analysis on Italian young adults. *Frontiers in Psychology, 11.*

Polizzi, C., Lynn, S. J., & Perry, A. (2020). Stress and coping in the time of COVID-19: Pathways to resilience and recovery. *Clinical Neuropsychiatry, 17*(2).

Reger, M. A., Stanley, I. H., & Joiner, T. E. (2020). Suicide mortality and coronavirus disease 2019—A perfect storm? *JAMA Psychiatry.*

Rohman, M., Marji, D. A. S., Sugandi, R. M., & Nurhadi, D. (2020). Online learning in higher education during covid-19 pandemic: Students' perceptions. *Journal of Talent Development and Excellence, 12*(2s), 3644–3651.

Serafini, G., Parmigiani, B., Amerio, A., Aguglia, A., Sher, L., & Amore, M. (2020). The psychological impact of COVID-19 on the mental health in the general population. *QJM: An International Journal of Medicine, 113*(8), 531–537.

Sica, L. S., Aleni, S. L., Syed, M., & McLean, K. (2018). I became adult when pathways of identity resolution and adulthood transition in Italian freshmen's narratives. *Identity, 18*(3), 159–177.

Trzebiński, J., Cabański, M., & Czarnecka, J. Z. (2020). Reaction to the COVID-19 pandemic: The influence of meaning in life, life satisfaction, and assumptions on world orderliness and positivity. *Journal of Loss and Trauma, 25*(6–7), 544–557.

Umberson, D., Crosnoe, R., & Reczek, C. (2010). Social relationships and health behavior across the life course. *Annual Review of Sociology, 36*, 139–157.

Wang, H., Xia, Q., Xiong, Z., Li, Z., Xiang, W., Yuan, Y., Liu, Y., & Li, Z. (2020). The psychological distress and coping styles in the early stages of the 2019 coronavirus disease (COVID-19) epidemic in the general mainland Chinese population: A web-based survey. *Plos One, 15*(5), e0233410.

Zarrett, N., & Eccles, J. (2006). The passage to adulthood: Challenges of late adolescence. *New Directions for Youth Development, 2006*(111), 13–28.

Multiple Systems in the *Meso* Domain: A Study in Organizational Cognition

Davide Secchi, Rasmus Gahrn-Andersen, Maria S. Festila, and Martin Neumann

Abstract In their book *The Theory of Social Organizing*, (Secchi et al., Organizational cognition: The theory of social organizing. Taylor & Francis, 2022) present a view of organizational cognition as it structures around *meso* domains. The *meso* is separate but intertwined with a *micro* (personal, individual) and a *macro* (structural) dimension. It is where social interactions allow for "social organizing" to arise. If we take this conceptualization seriously, it becomes immediately apparent that the structuration of the *meso* domain cannot be singular, but it requires multiple mechanisms to be in place at once. In this chapter, we argue that cognition can be framed as a system which integrates multiple domains, whose boundaries are plastic, and which is temporarily observable while leaving latent traces. In describing the above, this chapter then uses the example of the IOP 2.1.2 Model—an agent-based computational simulation—to explore what kind of systemic interdependencies exist among different *meso* domains.

1 Introduction

The study of cognition has traditionally been characterized by an emphasis on human individuality and, more specifically, on individual brain functions (e.g., Secchi & Adamsen, 2017). In the last few decades however, alternative positions have recognized the importance of a variety of resources that make cognition possible

D. Secchi (✉) · R. Gahrn-Andersen · M. S. Festila · M. Neumann
Department of Language, Culture, History and Communication, Faculties of Humanities, University of Southern Denmark, Søndre Stationsvej 28, 4200 Slagelse, Denmark
e-mail: secchi@sdu.dk

R. Gahrn-Andersen
e-mail: rga@sdu.dk

M. S. Festila
e-mail: marfes@sdu.dk

M. Neumann
e-mail: martneum@sdu.dk

G. Minati and M. Pietronilla Penna (eds.), *Multiple Systems*, Contributions to Management Science, https://doi.org/10.1007/978-3-031-44685-6_17

(since Hutchins, 1995; Varela et al., 1991). In breach of more traditional cognitive science, they postulate that the interplay between internal (brain-bound) and external (environment-bound) resources (Magnani, 2007) is essentially constitutive of cognitive phenomena in humans, which unfold in relation to socio-material practices. However, it is only in more recent times that there has been a call for the appreciation of the interdependencies that exist among the resources associated with cognition (Cowley & Vallée-Tourangeau, 2017). This essentially means that the different parts develop, interact, and co-evolve in such a way that they should be considered as constitutive of the cognitive as a whole. This point has several implications reverberating on our understanding of cognition, including that cognition is not tied solely to brain-states. At the same time, it also raises a series of questions that should be addressed in order for our understanding to reach an appropriate level of clarity.

First of all, it is unclear whether the claim that cognition is systemic refers to a fundamental "fact"—i.e., that cognition is a system—or to the way in which cognition should/could be studied—we can call this the *as if* hypothesis. Second, when discussing systems, it is essential to indicate which elements are included and which ones are excluded from them. In other words, it is sensible to discuss the boundaries of the system (formation, behavior, and disappearance) including its interrelations with other systems. Third, there is the question of time. If we assume that cognition unfolds as a system, then we shall discuss its evolution over time. In particular, when cognition is concerned, it seems that the system is temporary, although some of its parts may persist after the system has disappeared. Importantly, these surviving parts continue to be imbued with functional traces of the vanished system and they can become part of (new) cognitive systems. This ambivalence is particularly relevant to discuss since it is possibly a characteristic of a peculiar type of system—the cognitive system.

This chapter is organized around the three topics mentioned above and it uses cognition in organizations as a case in point. This is done to limit the scope of the discussion and, at the same time, to present a distinct set of cognitive activities that are probably among the most frequent for humans. The next section (Sect. 2) introduces the way in which we conceptualize organizational cognition by presenting the 3M model (Secchi & Cowley, 2021). The following section, Sect. 3, addresses the question of the *as if* hypothesis. Section 4 explores the main components of a cognitive system, while Sect. 5 is concerned with time. Section 6 discusses implications using a computational model as an exemplar and then offers a few concluding remarks.

2 A Theoretical Framework for Organizational Cognition

Secchi and Cowley (2021) introduce the 3M model as a means for re-orienting the discourse on organizational cognition around a systemic perspective. In this framework, individual (micro) aspects recombine with each other and with structural (macro) elements thus generating an inter-liminal space called the *meso domain*. It is within this domain that cognition unfolds because, as a relational phenomenon, cognition requires a space to connect the different elements of the micro and the macro together. For example, a Monday meeting where members of a team discuss the plan for the week to come is necessarily bound to the standard practices the organization uses to set meetings of this kind (macro structures). It is also set and takes place according to the infrastructure—both software and hardware—that the organization makes available to their employees and management (macro structures). The individuals who take part to the meeting, using specific resources available to them (micro), are also essential parts for cognitive activities to happen. As the emphasis on the *meso* suggests, neither micro elements alone nor macro aspects alone are sufficient for cognition to manifest itself. It is their dynamic combination and re-combination that sets the system. And Secchi and Cowley (2021) claim that these dynamics—e.g., sharing ideas, writing on a pad, interpreting a cue, debating a topic—emerge in the *meso* domain. Since most—if not all—of the meaning making that spurs off an organization's activities is anchored to existing meaning and/or interactions between individuals, the *meso* domain can be intended as a manifestation of *social organizing*.

In recent work, Secchi et al. (2022) draw on the initial conceptualization and write, as they describe cognition as *social organizing*:

> This is a perspective that postulates that cognition can only be understood when one is willing to consider it not just as dependent on the (micro) biological characteristics of individuals or on reactions to (macro) structural elements. Rather, these two need to be related to the way in which they are understood, individually and collectively, by the localized set of experiences, exchanges, interactions, exploitations, usages, communications, etc. In other words, it is how social resources (mainly human beings) enter cognitive activities that matters to us. If one reflects on any organizational practice, this claim seems self-evident. (Gahrn-Andersen et al., 2022, p. 6)

The above is relevant for our enquiry on systemic cognition because it takes the *meso* domain of social organizing as a starting point. Since this is a way for the other two aspects (micro and macro) to come to life in organizational cognition, we are by no means excluding them, but simply aim at finding more appropriate roles in relation to the system. In the following sections, we use the *meso* domain as the natural playground for human practice-based cognition.

3 The *as if* Hypothesis

The question above—whether the "systemic" attribute is a factual connotation of cognition or whether it is to be taken *as if* it shows those features—has an epistemological flavor. In fact, the actual question can probably be turned into one on systems that is on whether systems actually exist or whether they are a human construct used to interpret a given phenomenon. The problem can be framed as the question of whether ontology is a result of epistemology, thus assuming that a system stands in relation to the observer. However, if the question is intended this way, there is not much to argue. The point would be similar to asking whether a mathematical formula is an actual structure or just a way to represent a given phenomenon, being, or event. It is the argument that reflects the idea of the inner structure of events, objects, else. Confounding the representation (the epistemological tool) with the actual phenomenon would be a distortion, not only because it conflates the two into one, but especially because it suppresses the role of the observer. For this reason, the question is to be intended to depict cognition as a system or as something that shows systemic features but is not necessarily a system itself.

If we intend the above correctly, then we want to ask whether cognition is a system or if it works *as if* it is one. In the former case, there is an analogical reasoning in place, where all features of systems can (or should) be applied to the study of cognition. This means that we would need to rigorously define the elements that make cognition as well as their inter-relationships in a way that is sufficient to describe the workings of the system. In the latter case, this would not be necessary because we would only imply that there are properties of systems at play, but that it would be very difficult to isolate all of the composing elements. Taken in this view, the composing elements have to be understood not as properties of the system in isolation, but as relational properties of the system and the observer. A system itself would have to be understood as an element of the system-observer system.

Most proponents of the embodied, distributed, and extended (EDEC) view of cognition seem to propose some version of a system. For example, Hutchins (2014) is very much open about this perspective when he refers to cultural ecosystems. Proponents of radical embodied cognitive science (e.g., Chemero, 2009) explicitly refer to System Dynamics as their preferred representational tool. Instead, Cowley and Vallée-Tourangeau (2017) do not make any specific claim on systems; instead, they suggest that a "systemic view" is a way to *democratize*, to some extent, the role that the different parts play in cognition. This indicates that both the brain and the individual cognizer lose their centrality and become part of a more complex system of resources/elements, where relationality is more important than any of the single nodes of the network and, consequently, the system is fundamentally decentralized (cf. Gahrn-Andersen et al. this volume). This is probably closer to a perspective that takes the form of *complex adaptive systems* (Miller & Page, 2007). Seen from this perspective, cognition can be thought of as a system (see, e.g., Secchi & Cowley, 2018) comprised by multiple systems, one which is unpredictable, made of non-trivial interrelations among its parts, and which constantly changes and co-functions with other systems.

In light of this, it is probably safe to assume that researchers have used the system analogy multiple times before and that they have done so with some degree of success. Yet, as far as our knowledge is concerned, there have been few attempts to define a cognitive *system* in its basic components (see Gahrn-Andersen, Festila, and Secchi, this volume). This is what the following section is about.

4 Elements of Systemic Cognition

The highly variable set of circumstances that identifies systemic cognition makes it very difficult to find general abstract patterns one can follow. One such attempt has been conducted in Gahrn-Andersen, Festila, and Secchi (this volume) where generalities are expressed in terms of multiple ontologies. The focus of the present chapter is slightly different in that we are concerned with the boundaries of a cognitive system understood specifically in relation to the 3M model.

Here, reflecting on organizational cognition comes to help. The first consideration is about what we have called the *meso* domain. Since the claim is that cognition happens in this domain, where does it end? In other words, which elements characterize the *meso* and map its boundaries?

In order to take this point of enquiry, it probably makes sense to start from the model. A revisitation of it—a variation of the original Secchi and Cowley (2021)—is presented in Fig. 1. The three domains are intertwined in a way such that the Venn diagram shows overlaps between them. Both individuals and the macro superstructure contribute to making the meso possible while also affecting each other. The *meso* is constituted by impinging upon micro and macro elements at once, whenever they become necessary. But the way in which the 3M model is represented seems to suggest that there are three systems at play, each belonging to one of the Ms. Under this lens, the *meso* domain would be a system separate from the other two; even if separate, a case can be made to suggest that it leans upon the other two. Taking a more critical approach, however, it can be noted that the *meso* cannot exist without a micro and/or a macro domain. It is exactly this combination of cognitive resources that makes social organizing possible. Think of a simple task, that of writing an email to a colleague. The *meso* would be constituted by a series of structural elements, some material—such as a software, the office (if any)—others immaterial—e.g., practices, routines, procedural elements. It would also be made of the individual who writes the email and of his/her anticipation mechanisms—e.g., the email is written thinking of how it is going to be read, interpreted, and understood. The text itself—as one writes it—is an external resource that allows for cognition to attune itself to the task. In other words, it is *practice* that ultimately makes cognition what we observe it to be.

The example above serves the purpose of showing the strict interdependence between the three parts of the model. The system we are looking at is not confined to a *meso* domain, but it is composed by the entirety of the three Ms and the way in which they interact. By no means a focus on the *meso* means an arbitrary separation between the three domains.

Figure 1 presents the 3M model in a way that is slightly different from its original in that it shows multiple *meso* domains and multiple micro domains. The assumption here is that cognition is not univocally defined. Different individuals have different perceptions of the *meso* domain they contribute to define. In relation to the theoretical lens of different individuals, different *meso* domains come into being. At the same time, they may contribute to creating several parallel *meso* domains at once. Let us

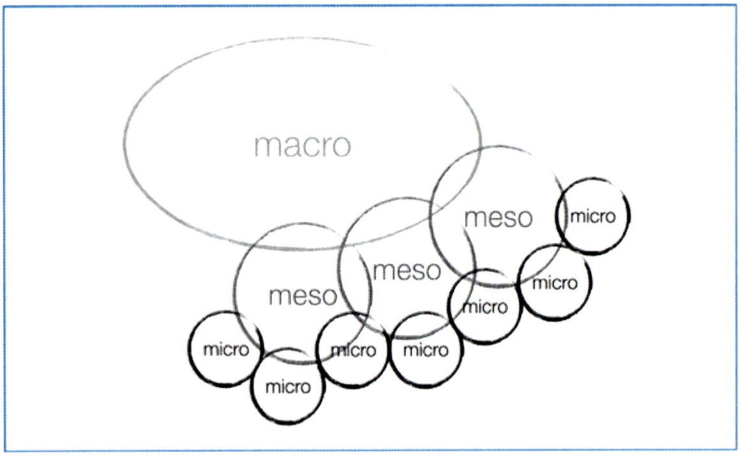

Fig. 1 The 3M model

go back to the example of the meeting. There may be a *meso* domain related to a slideshow a manager is presenting that is shared by the members of that meeting. At the same time, there is a separate domain that is still connected to the slideshow, but it relates to a text message one may send to a colleague outside of the room, to inform them of the content of the meeting or, perhaps, to express concern. There would be two separate (but connected) cognitive systems here, one related to the coupling mechanisms between resources in the meeting (Jensen et al., 2022) and one connected to a separate set of coupling mechanisms between one participant and someone outside of the meeting. If we consider the meso as the focus of cognition, then we have one system per each *meso* domain under observation.

As the examples above demonstrate, the boundary of the system can be very difficult to define ex ante because the constituents of a meso domain can vary quite significantly, depending on the context, practice, and situation. In very broad terms, we can tentatively state that organizational cognition needs all the Ms of the model to operate and the boundaries are moving—they expand or reduce—as a function of the environment. In an attempt to be more general, we can postulate the following:

$$m = \frac{f(M|T, i, \epsilon_M)}{g(\mu|T, i, \epsilon_\mu)}$$

where m is the meso domain that is a function of the macro (M) and the micro (μ) domains, T is time and i is the interconnections among parts, while ϵ is a component that may increase or reduce the effect of all the other variables in the system. Both M and μ are conditional on T, i, and ϵ. The two functions are such that the macro elements are factored in by the way in which micro elements unfold.

5 A System's Time

The third point that is relevant to understand when clarifying the systemic nature of cognition is the role of time. Much has been written about the timescales that affect such a system, especially in relation to fast and slow timescales that intervene on cognition at once (e.g., Cowley & Neumann, 2016). These are certainly relevant and play a different role depending on the domain of reference, as also recognized in the equation above. This section is concerned with another element that affects cognition as a system. This is the lifespan of the system.

The *meso* domains we have described so far can be all considered temporary. They typically have a start, one or a series of occurrences (happenings), and an end—which is very much what can be used to describe most systems. The peculiarity of cognition is that the lifespan is usually very short. Think, again, of the organization as a setting. Most (if not all) cognitive activities start and end in a matter of minutes (e.g., email, messaging), hours (e.g., meetings, production-related tasks), or days (e.g., hiring personnel, strategic planning). Yet, even when the activity is over, parts of it may leave a trace somewhere, either in the macro or in the micro domain. For example, repeated tasks may crystallize onto organizational routines (Pentland & Feldman, 2005) or individual habits (March, 1994). This is a process that has been referred to when the fast timescales feed the slow timescale, to some extent (Neumann & Cowley, 2016). However, this is not it. Some of these systems may leave traces that are more subtle, not easily recognizable, and that may remain latent for a long time, then re-emerge or get lost. Sometimes re-emergence is dictated by a specific configuration of resources that trigger specific behavior, or it could just be chance, improvisation, or serendipity (Bardone, 2011). On the one end, the system does end when the task, for example, is performed. On the other end, parts of it remain in a wider sphere (perhaps in a *meta-system*) and can become part of future systems. This implies that the lifetime of a system cannot be unequivocally defined.

This aspect of temporality—i.e., the system is temporary but its components remain available for reuse in some shape or form—is probably a source of ambiguity, especially when we refer to latent elements that cannot be directly observed, unless they become part of systemic cognition. A way to frame this could be that of using the extended cognition framework (e.g., Clark, 2003; Clark & Chalmers, 1998), specifically the reuse of resources that were part of previous *meso* domains. There is a level of dexterity associated with the repeated use of a resource that builds expertise, skills, and competences (Secchi, 2009). This may relate to the latent unobservable element of systemic cognition that has potentials to structure a *meso* domain.

6 Operationalization

An effective way to represent complex systems such as those related to cognition are agent-based models (ABMs). These are computational simulations that allow for heterogeneous and autonomous agents (e.g., cognitive resources) to interact in a given environment (e.g., a macro structural element where the *meso* happens) through a number of mechanisms or rules (see, e.g., Macal, 2016). ABMs have seldom been developed on EDEC perspectives, although they are a very powerful way to generate knowledge in these areas (as demonstrated in Secchi, 2021), specifically because of their ties to systems.

In this last section of the chapter, we use a model—the Intra-Organizational Plasticity Model, or IOP 2.1.2—to understand the aspects of systemic cognition reviewed above. The model has been introduced and first analyzed in Secchi (2020), and then extended and further explored in Secchi (2021, Chap. 11). The objective of the model is to understand what happens to task performance when waves of change invest an organization. A full description of IOP 2.1.2 can be found in the two publications mentioned above and in the OpenABM platform.[1] In the model, the employees seek colleagues to find how resources can help them perform their tasks. In so doing, they develop multiple *meso* domains (multiple systems), with various degrees of success—meaning that some organizing is likely to help them perform better. Each one of these *meso* domains is temporary, since they only serve the purpose of performing a task. However, when the task is performed, this may or may not leave a trace that is related to the operation already performed and that is unobservable unless called back actively. Let's call the latter *Condition A* and the former *Condition B* (Table 1). What the model shows is that two most common cognitive coping strategies emerge. A *differentiation* strategy is one that prevails under *Condition A*, and it is such that the variety of cognitive resources used (both social and non-social) allow for a wider reach and more "average" tasks can be performed. The other is a *concentration* strategy, one that emerges when *Condition B* is in place. This allows for a more specialized use of resources, function of a "stickiness" factor that comes into the system. In this case, the overall number of tasks performed does not increase much, while the most difficult ones are addressed more effectively.

The two conditions are symptomatic of the variability of the *meso* domain in that they show similar components—both in the macro structure (e.g., support for cooperation, resource availability, task structure) and in the micro domain (e.g., individual dispositions, connectivity, competence)—but result in fairly different cognitive strategies. Also, it is worth noting that the model produces several *meso* domains under the same condition, where the dominant strategy emerges as an average, not as something that every agent would necessarily perform.

The IOP 2.1.2 Model shows that multiple systems may co-exist at once, that their boundaries vary in relation to the conditions that set them up, and that they usually show limited temporality. In spite of that, there is a "carry-over" effect that structures different cognitive strategies.

[1] https://www.comses.net/codebases/4ff566a6-c0f8-4ca2-aa5f-21d44705aea8/releases/2.1.2/

Table 1 Comparing meso domains under two conditions

	Condition A (Use of resources as available)	Condition B (Repeated use of resources)
System boundaries	Wide; varied cognitive resources	Narrow; specialized use of cognitive resources
Time	The system leaves no trace, or traces are not actively called back, resulting in shorter timespan for task completion	The system leaves traces related to the operation already performed, and are actively called back, resulting in longer timespan for task completion
Cognitive coping strategy	Differentiation strategy (quantity of tasks solved takes priority over those that are more time consuming)	Concentration strategy (focus on the most relevant tasks)

Note This table refers to features and results of the IOP 2.1.2 Model as described in the text

7 Concluding Remarks

This chapter is an attempt to specify the conditions under which the *meso* domain can be used to define systemic aspects of cognition. The most sensible way to describe cognition is *as if* it were a system of elements that combine multiple domains in a *meso* domain making the boundaries extremely blurry. A limited lifespan with a latent unobservable tracing mechanism is also a relevant characteristic of this system.

Acknowledgements All authors acknowledge the funding awarded by the Velux Foundation of the project Determinants of Resilience in Organizational Networks (DRONe) (Grant Number 38917).

References

Bardone, E. (2011). *Seeking chances. From biased rationality to distributed cognition, volume 13 of cognitive systems monographs*. Springer.

Chemero, A. (2009). *Radical embodied cognitive science*. MIT Press.

Clark, A. (2003). *Natural-born cyborgs. Minds, technologies, and the future of human intelligence*. Oxford University Press.

Clark, A., & Chalmers, D. J. (1998). The extended mind. *Analysis, 58*, 7–19.

Cowley, S. J., & Vallée-Tourangeau, F. (Eds.). (2017). *Cognition beyond the brain. Computation, interactivity and human artifice* (2nd ed.). Springer.

Gahrn-Andersen, R., Secchi, D., & Cowley, S. J. (2022). Setting the scene. In D. Secchi, R. Gahrn-Andersen, & S. J. Cowley (Eds.), *Organizational cognition: The theory of social organizing* (pp. 1–15). Taylor & Francis.

Hutchins, E. (1995). *Cognition in the wild*. MIT Press.

Hutchins, E. (2014). The cultural ecosystem of human cognition. *Philosophical Psychology, 27*(1), 34–49.

Jensen, A., Secchi, D., & Jensen, T. W. (2022). A distributed framework for the study of organizational cognition in meetings. *Frontiers in Psychology*, 1560.

Macal, C. M. (2016). Everything you need to know about agent-based modelling and simulation. *Journal of Simulation, 10*, 144–156.
Magnani, L. (2007). *Morality in a technological world. Knowledge as a duty*. Cambridge University Press.
March, J. G. (1994). *Primer on decision making: How decisions happen*. Simon and Schuster.
Miller, J. H., & Page, S. E. (2007). *Complex adaptive systems. An introduction to computational models of social life*. Princeton University Press.
Neumann, M., & Cowley, S. J. (2016). Modelling social agency using diachronic cognition: Learning from the Mafia. In D. Secchi & M. Neumann (Eds.), *Agent-based simulation of organizational behavior. New frontiers of social science research* (pp. 289–310). Springer.
Pentland, B. T., & Feldman, M. S. (2005). Organizational routines as a unit of analysis. *Industrial and Corporate Change, 14*(5), 793–815.
Secchi, D., & Adamsen, B. (2017). Organizational cognition: A critical perspective on the theory in use. In S. J. Cowley & F. Vallee-Tourangeau (Eds.), *Cognition beyond the brain: Computation, interactivity and human artifice* (2nd ed., pp. 305–331). Springer.
Secchi, D., & Cowley, S. J. (2018). Modeling organizational cognition: The case of impact factor. *Journal of Artificial Societies and Social Simulation, 21*(1), 13.
Secchi, D., & Cowley, S. J. (2021). Cognition in organisations: What it is and how it works. *European Management Review, 18*(2), 79–92.
Secchi, D., Gahrn-Andersen, R., & Cowley, S. J. (Eds.). (2022). *Organizational cognition: The theory of social organizing*. Taylor & Francis.
Secchi, D. (2009). The cognitive side of social responsibility. *Journal of business ethics, 88*, 565–581.
Secchi, D. (2020). Cognitive attunement in the face of organizational plasticity. *Evidence-Based Human Resource Management, 9*(2), 192–208.
Secchi, D. (2021). *Computational organizational cognition: A study on thinking and action in organizations*. Emerald Group Publishing.
Varela, F. J., Thompson, E., & Rosch, E. (1991). *The embodied mind: Cognitive science and human experience*. MIT Press.

Multiple Clocks in Mental Time Processing

Fabrizia Giulia Garavaglia, Marco Giunti, and Simone Pinna

Abstract How does the mind keep track of time? Timing capacity is one of the key elements not only of learning, but also of our sense of identity and agency. This capacity is the most basic feature of the brain working, it connects an individual mind with its environment. May there be a metronome independent of other structures that acts as a unit of temporal measurement? There are two different possible answers to this question that imply a completely different understanding of the workings of attention, and so of individual consciousness. First, we could say that there is a specific area that always pulses at the same frequency, which works as a timer. In this sense, we have to suppose that it starts to pulse at some point of the development of the neural system. The features which define different contents of perception would then depend on those internal and subjective rhythms that are related to this original timing unit. Under this perspective, attentional content is a step-by-step sum of speedier activations compared to this timer. But, as we will argue here, this is only one possible answer, for we should also consider the existence of multiple clocks and an alternative way of measuring time might depend on their integration or overlap.

1 Introduction

This paper focuses on a timing-based level of explanation of cognitive processes that are involved in the learning process, in particular attention, memory, and conscious perception of agency. We will explore a new framework that tries to mediate between different perspectives, by overcoming their weaknesses but, at the same time, by

F. G. Garavaglia · M. Giunti (✉) · S. Pinna
Dip. di Pedagogia, Psicologia, Filosofia, Università di Cagliari, Cagliari, Italy
e-mail: giunti@unica.it

F. G. Garavaglia
e-mail: fabriziag.garavaglia@unica.it

S. Pinna
e-mail: simone.pinna@unica.it

G. Minati and M. Pietronilla Penna (eds.), *Multiple Systems*, Contributions to Management Science, https://doi.org/10.1007/978-3-031-44685-6_18

keeping their positive aspects. This will be achieved by focusing on a timing perspective.

In the first part of this paper, we will focus on informativeness, more specifically we will explore the notion of neural information, coding, and what we refer to when we talk of brain rhythms, neural spiking of a single cell or a group of cells, and the relevance of the timing component. Then, we will explore the relationship between neural activations and their timing coherence, and some of its cognitive implications, especially with respect to memory and attention. We will finally argue that causality attribution and agency ascription in individual learning, as well as the very concept of identity, all depend on a timing relationship.

1.1 Change in Timing Information. Learning and Conscious Agency

From a neurophysiological perspective, learning is explained by exploiting the concept of neural plasticity focused on morphological changes in the spatial architecture of different neurons through dendritic spines and the production of different synapses. This approach is mainly centered on the production, trade, and metabolism of neurotransmitters, receptors, and the enormous number of molecules that modulate the process. So, this kind of explanation focuses on material changes in neural connections at the molecular or cellular level.

Here, we delineate a different approach to neural explanation of learning processes. The focus will be on information, analyzed in its timing structure. The basic axiom that we assume here is that morphology is manipulated by information, which is codified in neural spiking. This means that learning is possible because of a previous change in the kind of coding and decoding in neural cells. What changes is the way in which neural cells codify information and send it to the cells that are connected to them. What is relevant for the occurrence of a learned output is consonance between different areas, or groups of neurons. And even the consonance between few cells could be truly relevant in interpreting the information coding of a piece of information. In this approach, the attention is not directed to spatial connections, but it is directed to pattern of activations and their timing coherence.

Under this perspective, learning is better explained as a change in neural pattern activity. Therefore, in order to better understand this process, we should explore the timing structure of the pulses of a neuron or a group of neurons. We can detect neural activations thanks to FMRI, MEG, or EEG, but we should analyze not only the part of the brain that is active under some conditions, but also the rhythm of neural dialog and its units of information or, so to speak, the alphabet of neural coding.

To understand learning processes, we need to pay attention to these changes in activation patterns. We will explore some discoveries in this research field, to better understand the relevance of timing, try to reinterpret learning processes under this perspective, and even give some hints on how to improve the outcome of these processes.

1.1.1 Coding of Intensity and Changing in Timing Information

Neural information is the root of neural plasticity, which determines learning processes. The neural communication code is generally described as the set of matching rules by which neurons represent the information that is received, processed, and transmitted. Neural communication has a spatial component that defines which cells are activated in response to a given stimulus, so it specifies across the brain areas that show a simultaneous activity with a specific stimulus. The neural communication code also has a temporal component (frequency codes and time codes), which describes the correspondence rules by which the electrical activity of a neuron encodes incoming signals and transmits the processed signal to subsequent cells. This second aspect of coding will be the focus of the present part. We will consider its philosophical implications for learning and agency perception in the second part of this paper.

The study of neural coding began with Edgar D. Adrian's discovery (Adrian, 1926) that the train frequency of action potentials generated in response to a sensory stimulus grows with the intensity of the stimulus. This suggests that the frequency of action potential constitutes a code that contains information about the stimulus, and more specifically about its intensity. The cellular basis of neural coding lies in the fact that each neuron receives synaptic inputs, distributed on a morphologically complex dendritic arborization, which are integrated at the level of the axon emergence from the soma, where the outcome of this synaptic summation over time determines whether and how many action potentials will be generated.

The study of neural coding aimed to describe the aspects of action potentials firing could be directly related to the various attributes of the stimulus: What is the information contained in that pulse train? What is the code used by neurons to transmit that information? The reverse problem, namely how it is possible to reconstruct a stimulus, is also studied by the examination of neuronal firing (the problem of neuronal decoding). These questions obviously have not a unique answer, and the theme of informativeness is very broad. Our target here is just to try to correlate some aspects of timing to attention.

1.2 From Intensity to Spike Timing and Attention

Despite all the debate about an exact definition of information coding in the neuronal system, it seems clear that the intensity of stimuli can be encoded in some way through the number of spikes in a time unit.

The same typology of stimulus hit the individual system with different levels of intensity, and this can be translated into the modulation of timing in the same neural areas. The proximity of height peaks in neural spiking is often observed to be coherent and synchronic with high intensity of the input, to which a given circuitry is subjected, whether low oscillatory frequencies seem to indicate low intensity of the stimulus. The timing of the neural activations conveys information about the stimulus and its processing can indicate something relevant also about the conscious content connected to that stimulation. We must also observe that intense stimuli involve our attention in a stronger way. Intense contents are more likely to become aware because they manifest more urgency, as they require a response.

This claim, about a necessary connection between intensity of stimuli and attention, could be challenged by some empirical evidence that shows that, under certain conditions, we could be completely blind to some very evident features of the environment. This is the case of many illusions, as a paradigmatic example, we could consider the *invisible gorilla* (Chabris & Simons, 2011), which shows everyday illusions governed by beliefs about the world in a top-down process. In this example, there is a strong claim about the fundamental role played by attention and the beliefs about world laws that seem to guide our perception.

The claim that our perception is illusory is quite old in the history of human thought. But our point, here, is to move further, beyond perception illusion and attention inaccuracy. We would like to question the very nature of the attention that is supposed to guide our conscious perceptions. From a phenomenological standpoint, whether our perception is false or illusory is not relevant. We do not want to investigate in depth the supposedly illusory nature of subjective perceptions of the intersubjective real world.

Our subjective perception, of either time or the world, is not illusory for ourselves. We think that, in order to understand conscious perception in learning, we should not forget or avoid considering this important and distinctive aspect.

In particular, we wish to focus on the nature of attention, by exploring those morphological and timing aspects that can be relevant for its existence. We focus on attention and memory timing features because they are fundamental components of learning processes.

We will analyze more deeply the timing properties of information in attention processes, to try and define a timing-based paradigm that could help to differentiate between voluntary learning processes and involuntary ones, as well as among the different agency perceptions connected to them. The paradigm that we propose will incorporate some relevant aspects of epiphenomenal, predictive, and enactivist theories, by considering timing as the connection between subjective internal processes and external stimuli.

2 Timing in Learning

Before approaching attention and conscious perception from a timing perspective, we want to talk about some crucial timing features of learning processes. We want to highlight one very relevant aspect of neural plasticity that here could be relevant in this timing context: Spike-timing-dependent plasticity (STDP) in learning.

Hebbian reinforcement, which means the strengthening of synaptic junctions, has a specific time window in which it operates. Spike-timing-dependent plasticity is precisely what guarantees the causal interpretation because it refers to the sequence of activations in a synaptic bilateral conjunction (Recurrent Synapse).

If we consider two neurons connected to each other and one activates for a stimulus that occurs before, there will be a strengthening of the synapse in one direction (exactly the direction which starts with the neuron that activates with the first stimulus) and not in the opposite one. This is demonstrated in several studies and could offer a possible argument for the way in which the brain interprets causality, just like a sequence of temporal activations. It is also worth noting that this explanation is coherent with Hume's psychological theory of causality by Hume (1748). In his opinion in fact, causality is only the experience of repeated connections in time.

There is another important aspect that we should consider here: the time window necessary to activate this connection has a grain of milliseconds, in fact an interval of a second is not efficient anymore (Levy & Steward, 1983).

2.1 *Timing Contiguity and Ascription of Real Identity*

This observation leads us to approach another problem that is crucial for the attribution of identity and, probably, for the sense of subjective action. This is the problem of the so-called size–distance invariance. Starting at a very young age, humans are able to recognize the identity of the same object despite its changing appearances in shape and size due to the distance and angle of perspective. More or less, all the features and aspects of a thing always change depending on distance, circumstances, light condition, movements, or perspective. But still, we are always able to recognize the identity of a thing despite all these different appearances and it seems that we possess this ability since we have been infants.

So, even if this is not meant to be a conclusive answer, we would propose that timing contiguity is fundamental for the ascription of identity. The way in which the mind is able to recognize features that must be considered the same despite their different appearance, in contrast to features that should be distinguished, depends on timing.

Timing generates the ability to discriminate between different things in the environment. If some activation configurations always occur with a very close temporal contiguity, they are associated in a meaningful unity by the subject. Individual body involvement in perception is discussed in Noë's work (Noë, 2004). Our present proposal can be seen as an extension of the argument which considers bodily timing involvement.

2.2 How Does the Brain Count Time?

Intense stimuli are represented at the timing level (in the sense that a stronger stimulus is represented by a higher frequency than the one that represents a weaker stimulus), and they have a stronger impact. It seems that attention (and conscious perception) is the result of a process of comparison between different stimuli through a proper timing neural structure, where the speediest activations seem to be the most implicated in attentional contents. Is there in the neural system a structure that acts as a metronome? May there be an internal clock that gives a unit of time to the system? A metronome independent of other structures that acts as a unit of temporal measurement and makes it possible to select high speed signals, which are more urgent with respect to others and so have the dignity to become conscious?

There are two different possible answers to this question that imply a completely different understanding of the workings of attention, and so of individual consciousness. First, we could say that there is a specific area that always pulses at the same frequency, which works as a timer (Gibbon et al., 1988). In this sense, we have to suppose that it starts to pulse at some point of the development of the neural system. The features which define different contents of perception would then depend on those internal and subjective rhythms that are related to this original timing unit. Under this perspective, attentional content is a step-by-step sum of speedier activations compared to this timer. But, as we will argue in the next section, this is only one possible answer, for we should also consider the existence of multiple clocks and an alternative way of measuring time might depend on their integration or overlap.

There are several reasons why the mind should be able to detect and measure time. We need this to explain our understanding of synchronic perceptions that are not synchronic at all in the real intersubjective world. The ability of the brain to measure time is also needed for a simpler job like understanding the provenance of a sound[1] or for interpreting a sentence thanks to the capacity to correctly separate syllables and words.

So, how can the brain measure time? Is there a specific area that is structured for this, or are there multiple interconnected areas that do this complex and fundamental job? And what is the relevance of these observations for distinguishing different learning processes and different senses of individual agency? The next section aims to answer these questions.

[1] The brain must calculate, in a very short time, the distance that separates our ears, to be able to understand exactly where sound waves hit the body, and so where the sound comes from.

2.3 Multiple Clocks and Consonance

This section aims to make clear that it is exactly the timing capacity that is one of the key elements not only of learning, but also of our sense of identity and agency. This capacity is the most basic feature of the brain working, it connects an individual mind with its environment, it is crucial for the sharing function of the workspace (Dehaene & Naddaché, 2001) and for predictive and comparing activities.

One of the most relevant rhythms in our life is the circadian one. When we talk about this, we refer to the regulation of the individual that determines a period of sleep and a period of activity. Surprisingly enough, many experiments demonstrate that it does not only depend on the environment light conditions. In fact, even if the individual is in a condition of artificially regulated light, or even in complete darkness, she is able to preserve a circadian rhythm (Lavie, 2001). This makes us think of a specific area that is specialized in measuring time.

The suprachiasmatic nucleus was considered one of the best candidates, because when it was transplanted from an OGM rat (that was genetically modified to have a circadian period shorter than the normal one) to another normal rat, produced the same circadian rhythm in the new brain (Ralph, 1990). But in humans, this is not enough. It is not true that the suprachiasmatic nucleus is able to regulate all the other rhythms with a finer temporal grain. In fact, we need to measure different time periods and durations, depending on cognitive activities that process different stimuli with different timings.

If there are so-called *place cells* that activate when we occupy different regions of space, it seems that there are not analogue cells for time. But it is true that there are so-called *clock cells,* especially the cells of the suprachiasmatic nucleus, that exhibit a proper inner rhythm, in the sense that they cyclically realize different proteins and neurotransmitters (Reddy, 2006). This mechanism which regulates our circadian rhythm is known as Transcription Translation Autoregolatory Feedback Loop and it is organized through to the modulation levels of the protein named Period (Kanopka & Benzer, 1971). However, it is not true that they respond to stimuli in a cyclical way.

Moreover, we need to highlight a very important aspect of time measuring. Circadian rhythms organize events in the body (like the production of hormones and proteins) that are shorter than the circadian period. There is an infra-periodic measure and an over-periodic measure. The first measure is used to organize events that are shorter than the period. The circadian clock, in fact, has a period of 24 h and, with respect to it, hormones and molecules are produced in accordance with a shorter period (many times in 24 h). The second measure is used to count and organize events that are longer than the period.

Drug abuse studies and neuropharmacology are also useful to investigate clockwise functioning in the brain and also time perception. These substances are responsible for changes in neural activity because they have a mimic activity. This means that they act like endogenous neurotransmitters altering the transmission of information in a specific network for example, sometimes preventing the reuptake of some molecules, sometimes stimulating the abnormous production of some molecules.

They also often have an impact on time perception, but this may be a misleading observation.

On the one hand, saying that time perception is altered implies the existence of an internal normal perception; this in turn presupposes an internal normal clock with respect to which perception of a duration turns out to be altered. But this is not a good argument, first because alteration of time perception strongly depends on the subject and the circumstances in which the drug assumption occurs, and second because there could be the same temporal alteration (for instance the perception that time passes very fast) with drugs that act on the individual system in opposite ways, like cocaine (stimulant) or marijuana (relaxing).

Moreover, some experiments testify that we are able to exercise, and so to improve, the ability to evaluate a given time duration without improving the ability to guess longer durations. This means that through the repetition of the same task, we become, step by step, more precise in estimating the duration of an experience that has a specific time window of duration, but we fail when we are supposed to estimate a thinner or bigger window even of that very experience. We need some systems that are sensible to very fine time scales and others that are sensible to coarser time scales.

All these findings induce us to think about an alternative explanation of timing activities that takes into account the multiplicity of relevant time windows. We should thus consider the existence of multiple clocks in the brain that are sensible to different features and granularities of time, can perform different activities, and might be connected to the nature of our cognitive perception of time and agency. We could try to figure out the work of different clocks in the brain through different clepsydrae that count different time portions. We could have different clocks that are sensible to different time grain that the brain uses to organize different activities, but in some specific moments, they are coherent between them.

We can think, in fact, of many periodic oscillators with different frequencies that coincide on some of their peaks, so that their partial consonance forms a periodic structure. In this case, we do not have a pulse that functions as a measuring unit, but the measurement is due to the consonance between different clocks. This means that time is processed through a population code in which every neuron plays a distinctive role in the global dynamics of the system (Fig. 1).

2.4 Timing and Conscious Agency Perception in Learning

Timing sequences could also be the root of different agency perceptions in different motor learning procedures. When we started this work, the question that we moved from was about the difference that we can detect in voluntary (active) or afferent (passive) learning, despite in either case neural changes are more or less the same.

If we consider the learning process of a motor action, we see in either case neural changes due to neural plasticity that can be explained by Hebbian learning and STDP.

When we learn something new through a voluntary process, we have a strong perception of individual agency. At the same time, a modification in our brain occurs,

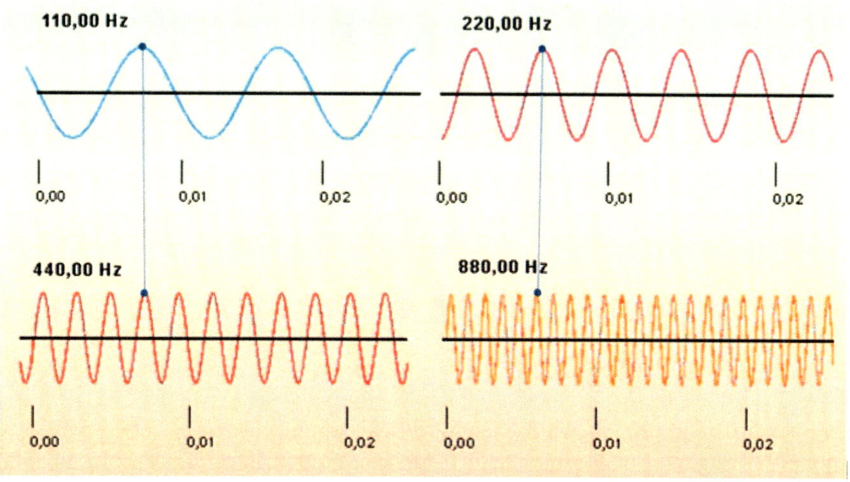

Fig. 1 This example is taken by music harmony, but it is useful to visualize different clocks in the brain that show a coherent peak at some points (in this figure such points correspond to 0.006—blue lines—and some of its multiples)

due to neural plasticity, that could represent the effect of the reproduction of the action. But we can obtain a strong modification in neural pathways also in the case of passive reproduction of a motor action through the use of prostheses, orthoses or other devices. Some of these tools limit the degrees of freedom of action movements, and so are able to strengthen the networks that correspond to specific and effective movements (Pittaccio et al., 2013).

Other kinds of tools aim to improve learning through a passive stimulation of a peripheral configuration of body parts. An example is a robotic glove that touches sequentially and repeatedly a sequence of fingers, helping the subject to perform a song which corresponds to that sequence of fingers on the piano keyboard. It is important to stress that, in both these cases, the subjective sense of agency is not vivid as in the active learning cases.

Obviously, it is not possible to experimentally compare exactly the same areas and the changes in neural plasticity that are the effects of voluntary learning on the one side, or afferent learning on the other side. First, because of the enormous number of connections. Second, because experimental conditions to compare these two different ways of learning require two different subjects and, consequently, two different brains, or the same subject, but with the involvement of her two different hemispheres.

Nevertheless, if we consider many studies on neurophysiological, morphological and temporal observations, we could try to delineate a timing-based explanation for the different senses of agency related to these two different kinds of learning.

In the case of voluntary learning, the causal process seems to reproduce the pathway of STDP, as the speed of a neural configuration related to a specific motor task becomes faster and faster with repetitions. In the case of conditioned learning, even if we want to reach the target of a learned output, the sense of agency is compromised by a different timing configuration. The difference is neither in the motor part that is active in the neural individual system nor in spatially different connections. The reason that justifies different senses of agency might instead be primarily linked to different timing, or more precisely, to the way in which multiple clocks integrate each other.

3 Conclusions

The approach we have presented is a theory of timing capacity where, even if neural components are important, what is fundamental is timing coherence between internal multiple clocks. This is also modulated by the temporal architecture of environmental stimuli. The relevance of the environment depends on timing proportionality and coherence, and not only on its material components.

We have argued for the relationship between timing architecture of neural activations and some of its cognitive implications, especially with respect to memory and attention in learning processes. We have discussed coding of stimulus intensity as change in timing information, and we have correlated timing contiguity with ascription of identity. The observations we have examined imply the existence of multiple clocks which sustain this consonance activity.

Our view is mostly focused on timing architectures. Potentially, the same information could be processed in completely different brain locations if the temporal structure of the activations is preserved.

Under our perspective, conscious perception is a matter of timing organization and it depends on the way in which different external stimuli, with proper temporal structure, interfere and are processed by different clocks in the neural system. These multiple clocks are able to change information, and this determines further changes in the whole individual body, which is the interface or the medium of interconnection between external stimuli and the individual system.

The consonance in internal structures in the subject generates a change in the organization of multiple clockwise structures, and this interconnection may be seen as the onset of the conscious perception of identity and agency in performing an action.

References

Adrian, E. D. (1926). The impulses produced by sensory nerve endings. *The Journal of physiology, 61*, 49–72.

Chabris, C., & Simons, D. (2011). *The invisible gorilla. How our intuitions deceive us.* Harmony Books.

Dehaene, S., & Naddachè, L. (2001). Towards a cognitive neuroscience of consciousness: Basic evidence and a workspace framework. *Cognition, 79*, 1–37.
Gibbon, J., et al. (1988). Scalar expectancy theory and choice between delayed rewards. *Psychological Review, 95*, 102–114.
Hume, D. (1748). *An enquiry concerning human understanding*. Trad. it. di Dal Prà M., *Ricerca sull'Intelletto umano*. Laterza 2009.
Konopka, R. J., & Benzer, S. (1971). Clock mutants of Drosophila melanogaster. *Proceedings of the National Academy of Sciences, 68*, 2112–2116.
Lavie, P. (2001). Sleep-wake as a biological rhythm. *Annual Review of Psychology, 52*, 277–303.
Levy, W. B., & Steward, O. (1983). Temporal contiguity requirements for long-term as sociative potentiation/depression in the hippocampus. *Neuroscience, 8*(4), 791–797.
Noë, A. (2004). *Action in perception*. MIT Press.
Pittaccio, S., et al. (2013). Passive ankle dorsiflexion by an automated device and the reactivity of the motor cortical network. In *5th annual international conference of the IEEE engineering in medicine and biology society (EMBC)* (pp. 6353–6356).
Ralph, M. R., Foster R. G., & Davis F. C. (1990). Transplanted suprachiasmatic nucleus determines circadian period. *Science, 247*(4945), 975–978.
Reddy, A. B., et al. (2006). Synchronization and maintenance of timekeeping in suprachiasmatic circadian clock cells by neuropeptidergic signaling current. *Biology, 16*, 599–605.

The Interpsychic: A Leading Actor in Interpersonal Relationships and in the Psychoanalytic Scene

Elisabetta Marchiori

Abstract Freud's scientific model of reference is related to a linear, deterministic rationality, which adopts a homeostatic, stable system concept. Over the last few decades, Psychoanalysis, like all sciences, has approached the theory of Complex Systems, recognising its strong innovative impact for the study and understanding not only of psychic phenomena in the relationship between mind and body, but also of what happens in the psychoanalytic process. The aim of this contribution is to present the concept of the interpsychic, proposed by the Italian psychoanalyst Stefano Bolognini. This notion can be considered of some relevance with a view to contributing to a reflection on possible convergences between Psychoanalysis and Complex Systems. This concept, developed on the basis of clinical observation of phenomena occurring during psychoanalytic treatment, concerns all human relationships and refers to fundamental physiological processes. In addition, possibilities are being explored to extend the research and clinical use to other areas, such as the developmental age, groups and institutions. This topic undoubtedly requires a more articulated theoretical reflection, which is beyond the scope of this work. It aims to show, through a clinical case and with evocative language, how the interpsychic can be considered a significant protagonist or agent of change in the analytical scene, where complex dynamic interactions, even unpredictable ones, take place.

1 Introduction

The Freudian scientific reference model, linked to a linear and deterministic rationality, has evolved into the more recent clinical–theoretical models of 'field theory', object relations, intersubjectivist and interpersonal approaches up to Bucci's 'multiple code theory' (2021). They seem to take the theoretical references of Complexity Theory for granted (Lenti, 2014).

E. Marchiori (✉)
Italian Psychoanalytic Society and International Psychoanalytic Association, Padua, Italy
e-mail: lisbeth.marchiori@gmail.com
URL: https://www.elisabettamarchiori.it

G. Minati and M. Pietronilla Penna (eds.), *Multiple Systems*, Contributions to Management Science, https://doi.org/10.1007/978-3-031-44685-6_19

De Robertis, in her paper 'The Logic of Complex Systems: a Potential for Psychoanalytic Theory and Clinic' (2005) shows how the concept of System provided Psychoanalysis, with a new theoretical and epistemic support for relational and interactive factors, challenged the theoretical framework of classical metapsychology. Psychoanalysis was thus able to turn its attention 'to the subjectivity of the psychic understood as a dyadic, interactive, bipersonal system in the frame of mutual regulation' (ibid., 1–2). This author highlights how the concept of the System used by Freud is homeostatic, stable and does not accord with the dynamism of biological systems. In fact, the stimulus disturbs its configuration and tends to be eliminated by the system itself.

Contemporary Psychoanalysis aims instead to explore open, complex, dynamic and non-linear Systems, with a view to studying psychic phenomena in evolutionary terms. In Complex Systems, the stimulus generates a rebalancing, which is the effect of a self-regulation aimed at a new equilibrium. The explanatory resources of Complex Systems are therefore important with respect to both theory and the clinic, in the perspective of treatment and possibility of change.

Lenti (2014, 2021), in his proposal to elaborate an epistemological model based on Complex Systems Theory applicable to Psychoanalysis, suggests that the system constituted by the analytic couple at work could be considered 'emergence'. Indicating the qualities or properties of a system, that presents a character of novelty with respect to the qualities or properties of the components considered separately. Indeed, it is well established that the phenomena that occur during the analytic process are co-constructed within the analyst–patient couple.

This author argues the analyst–patient relationship can be considered a 'systemic coupling' (Lenti, 2021, 32) when it fosters, through its own tools, evolutionary change.

This concise review of some contributions by Italian authors specifically dealing with the relationship between Complex Systems and Psychoanalysis is useful to introduce the concept of the interpsychic, proposed by Italian psychoanalyst Stefano Bolognini, which can be considered of some relevance in terms of contributing to a reflection in this field of research.

2 What Is the Interpsychic ?

In human relationships throughout existence, from the earliest stages of life, there is an innate tendency towards fusionality and at the same time separateness towards the other.

The concept of interpsychic proposed by Bolognini (2008, 2014, 2015, 2016a, 2016b, 2019, 2022) refers to a level of psychic functioning in which partial areas of physiological fusionality can form between two subjects, favouring preconscious exchanges, while maintaining intact the perception and awareness of separateness.

Bolognini led a theoretical–clinical study group around this concept between 2013 and 2020, in which around thirty psychoanalysts participated, meeting monthly. The group's work on clinical material referred to patients undergoing psychoanalytic treatment, psychotherapeutic treatment or followed by the Psychiatric Mental Health Services. The group's reflection work focussed in particular on how analyst and patient can be 'combined' in the working couple, creating interpsychic transitions. During the group's discussions emerged the possibility of extending the study of the interpsychic to broader areas, such as developmental age, group dynamics and institutions.

The group's meetings were suspended due to the pandemic, which provided the opportunity to write the book *L'elogio della gattoiola. Esplorazioni intorno all'interpsichico* (*The praise of the cat flap. Explorations around the interpsychic*, Marchiori et al., 2022a). It collects and deepens the reflections developed in the group over the years, also on issues of psychoanalytic technique and metapsychology. The title refers to one of the two metaphors Bolognini uses to illustrate the notion of the interpsychic. That of the 'cat flap' refers to that small opening in the doorway through which a cat can freely enter and leave the house. The passage of the 'cat flap' is configured as a 'secret passage' that facilitates internal flows between the conscious central Ego and the preconscious Ego and allows moments of physiological, occasional and non-confusional fusionality between the subjects. The 'cat flap' device topically corresponds to a preconscious mental level and an interpsychic relational level (Bolognini, 2008, 74).

The other metaphor is that of the 'porticos of Bologna' (Bolognini, 2019), characteristic of this city; they are covered passages, areas neither outside nor inside, intermediate between private and public space, where people enjoying protected sharing. The intermediate nature of the portico, where people can meet and exchange a few words without the need for official introductions, allows contacts to take place in a non-intrusive and mutually acceptable manner.

These two metaphorical images illustrate well the definition of interpsychic as it manifests itself in the relationship between two subjects: 'A mode of functioning that internally relates two individuals in a healthy, livable and economical manner. It is a highly permeable functional level shared between two psychic apparatuses' (Bolognini, 2022, 1). It is a physiological and adjustable modality, deeply connected to the area of transitionality, as it concerns the transition from Self to Non-Self.

It also characterises individual intrapsychic functioning, which can be found in the cooperative/relational organisation and style between the inner parts of the Self: it can be considered 'broadband and low energy', because it utilises pre-subjective and/or co-subjective elements harmoniously and naturally (Bolognini, 2022, 3).

In the dimension of the interpsychic, the work of the analyst–patient couple is based on those bodily equivalents that Freud had already noted in the *Three Essays on Sexual Theory* (1905), where he explains that being held, the act of sucking, phallic attitudes and creative genital encounters represent equivalents of imposed or consensual modes of relating, forced or enjoyed by both members of the analytic couple (Marchiori et al., 2022a, 2022b, V).

The interpsychic, which differs from the intersubjective and the interpersonal, is thus an occasional and natural mode of co-experience and cooperation that connects two individuals. As a prototype of such a connection can be considered that between a mother who breastfeeds her infant, where the natural cooperation between mouth and nipple allows the passage of internal, concrete and affective contents between the two. This can be considered the model for the later stages of relationships between individuals, including the analytical one (Bolognini, 2022).

In summary, the interpsychic can therefore be considered the psychic equivalent:

(1) of the healthy and necessary fusion conditions and processes that in nature enable vital exchanges (of nourishment, feeding, regulation) between human beings in the early stages of life and the following development;
(2) of the intimate relationships that at all stages of life, including adulthood, allow the natural and non-pathological passage of content from within a subject to within another subject: in sexuality, in learning, in the various areas of development;
(3) of the mucous membranes that allow contact between surfaces belonging to different subjects, since they are located in the transition zones between the outside and the inside of the human body; they can facilitate, by moistening, or hinder, by drying and hardening, the passage of substances or parts of the body through the intimate junctions.

3 Why Is It Important to Recognise the Interpsychic Relational Level?

Generally speaking, during each session with the patient it is crucial for the analyst to identify the level of the relationship—with its related phantasmatic configurations—that allows them to understand and make the best use of what is happening, what kind of exchange is taking place, which modes of contact may possibly favour more adequate progress and which bodily equivalents of reciprocal combination are operating unconsciously.

It is particularly important to be able to recognise those transformative situations in which the subject's Ego (depicted as the 'landlord') can allow itself to lower its defences and can trust the autonomous movements of the preconscious (depicted as the 'cat'), through the interpsychic passages (depicted as the 'cat flap') thus creating a potentially transformative interpsychic exchange (Bolognini, 2022).

In this regard, Lenti (2021, 32), in his perspective, also states that 'monitoring the state of the patient–analyst system must be carried out constantly. Indeed, evolutionary leaps in a complex system highlight the unpredictable nature of emerging forms and evolutionary bifurcations can be utilised by introducing information, for example, through interpretations'.

It is interesting to emphasise that interpsychic happenings are not predictable or actively induced by some specific intervention of the analyst. One can recognise and use them as agents of change when they manifest themselves as an 'unexpected' result of a combination of elements that occurs during a session. In those moments, an interaction takes place in which the contents of one subject moves to the other at a pre-subjective level, according to physiological modes of functioning (Battaglia et al., 2022, 20–21).

I, therefore, report a useful clinical example to describe the phenomenology of the interpsychic during the psychoanalytic process. In order to clearly and effectively show and frame the dynamics between analyst and patient, I have resorted to cinematographic language which, due to its metaphorical value, can facilitate the intuitive understanding of what can be considered interpsychic passages. One enters the analytic scene with overviews and close-ups, tracking shots forwards and backwards, in order to have the possibility of looking at the clinical material together with the reader/viewer: single frames, scenes and sequences, short and full-length films related to the interpsychic and its psychic equivalents of interbody processes (Lagrasta et al., 2022).

4 A Clinical Example: *Pas de deux* (duet)[1]

Leda is a young woman with a petite and graceful figure, who has always dreamed of being a dancer. When she first arrives at my consultation room, she has just graduated and is experiencing a deep crisis situation. She has decided to end the relationship with a man she was supposed to marry and is thinking of abandoning her professional career to devote herself exclusively to dance, her great passion.

We began psychoanalytic treatment together that lasted several years with three weekly sessions.

Leda is the youngest of four children. Her father, who had devoted his life to his job and his collection of war relics, is described as an *'introverted, rigid man, incapable of relating to others'*. Her mother, a housewife, is portrayed as a woman *'sacrificed to the family'*, always anxious, depressed, short-tempered and busy: *'For her, everything translated into drama and catastrophe'*.

I think of a couple of parents who, rather than 'feeling and being parents', 'did the parenting' and that Leda, instead of living 'to be', lived 'to do'. Indeed, she says: *'But if my parents were so tired, they shouldn't have had all these children … and if they really didn't want me, they could have put me up for adoption!'*.

I hypothesise Leda's desire to be *adopted* by an analyst in whom she could find, in the transference dynamic, the parental couple with whom she could feel desired and recognised.

Leda, over the years, has danced a number of choreographies on the analytical scene, from the simplest and most repetitive to the most complex and swirling. Her repertoire has been demanding and has involved heavy psychoanalytic work, aimed at integrating the dissociated parts of the Self, between 'doing and saying' (Racalbuto, 1994), between 'doing and being', between body and mind, passing through the vicissitudes of sexuality and Oedipus.

[1] This clinical case of mine has already been published in Lagrasta et al. (2022), I thanks the chapter's co-authors and the publisher Alpes Italia for permission to publish a revised version here.

In a series of flashbacks, we see Leda dancing, with great passion and commitment, as a child and then as a teenager. Her parents are not among the spectators, they have never supported her in the development of her innate talent, but boycotted her, fearing she would neglect her studies. So she had given up dancing, developing an eating disorder. Her parents had not worried about it, having ascertained that '*she was not pregnant and not on drugs*'. Through behavioural therapy, she had laboriously reduced her symptoms so that she could study and work to pay for her dance lessons, but the deep suffering remained.

The framing returns to the present, the scene takes place in the consultation room, when Leda is asking for help: '*I feel overwhelmed by the feeling that I have never stopped, I have no internal points of reference*'—she is saying—'*I don't know what to do with my life, I don't know who I am or how I could be*'.

In the first phase of analysis, I was a spectator of the images that emerged during the sessions. There were dream sequences mixed with those of fantasies and excerpts of real life, or sequences of stories that proceeded in parallel, with abrupt interruptions and temporal transitions, sometimes difficult sometimes difficult to link together.

I was able to assume a gradually more interpretive role when the scene, as the patient described, '*transformed from two-dimensional to three-dimensional*', creating a space that could contain both, her and me. It was thus possible to assemble the fragmented and disjointed sequences into a shared narrative with meaning. Leda's body also took shape and consistency, coming into contact with the internal and external world through gradual, albeit discontinuous, steps towards subjectification.

The defensive organisation oriented towards omnipotent fantasies of self-sufficiency had gradually loosened, allowing a kind of fluidification and permeabilisation of the mucous membranes, to use the metaphor of the psychic equivalents of inter-corporeal processes.

Leda's solo dance seems to have turned into a 'duet' together with the analyst. As Bucci (2021, 259) writes, the analytic exchange can be seen as a choreography: 'No one dances the tango alone; no one knows exactly what step will happen until it does; each dance comes alive in the interaction between the participants'.

At this stage the interpsychic enters the scene with increasingly important, full-bodied and defined parts. This can be recognised in this dream sequence, which Leda recounts when she begins to be able to think that her analysis will not be interminable and that the time has come to separate. It is a dream 'that turns the page' (Quinodoz, 2001) and stages the psychic functioning achieved: '*I had to come to the session and ... this is the strange thing ... it was taking place in a garden, we were both sitting on a bench. I felt in a state of inferiority, a bit childish compared to you and this gap initially made me uncomfortable. As I kept quiet, you said to me: I have learnt a lot from you, Leda. Yes, you called me by my name, I thought that my feeling of inferiority was not caused by you, it was my own that I now recognise. In the dream, you were a little moved and I too was a little moved ...* (she cries softly)... *and even now I am moved ... because then I felt relieved... then the cat woke me up from my sleep*'.

Together, we come to understand how, through the dream, Leda manifests her desire to feel her needs listened to and recognised for what she has become. She wants to 'separate' from her parents and her analyst, in a climate of sharing and no longer one of rejection or conflict. '*How much I would have preferred*'—she adds—'*that my parents spontaneously left me a minimum of autonomy, I always had to take it by force. For them I am still a child who has to do what they want, otherwise they won't accept me… I have always struggled to enter into a real relationship … because when you enter into a relationship with a person … you enter into a relationship!*'

'Yes, there's no doubt about it!', I answer her.

A glance forward frames Leda's face, who, laughing softly, confirms: '*Actually… I get it*!' She is almost surprised to grasp something that now seems obvious to her, but for a long time had not been.

In this session, I believe I resonated with Leda in the area of the interpsychic, dancing as a couple. We are in the dimension of that level of functioning of 'symmetry within asymmetry', of 'cooperative fusionality', which can be considered the equivalent of being on the same boat, travelling on the same bus, walking on the same porch, swimming in the same pool' (Bolognini, 2008). A situation similar to the one depicted in the dream, where we are sitting, Leda and I, on the same bench.

In this kind of analytic exchange, the subject can experience the other within the Self while finding new parts of the Self, as happens in the experience of dance.

The manic defences have diminished over time, the depressive anxieties reduced, the internal objects reclaimed and the relationships made more liveable. Leda has become the protagonist of a life that belongs to her, she is able to live the present and think about the future.

5 Conclusions

Through Leda's story, I have tried to show how the interpsychic can be considered a leading actor in analytic treatment and to focus on the role it can play in a transformative perspective.

The use of cinematographic language, associated with some metaphorical images of inter-corporeal equivalents, has been used to facilitate the intuitive understanding of this concept that concerns all interpersonal relationships, both individual and group. Specifically in the relationship between analyst and patient, it allows us to see it 'in action' (Hunter, 1994), triggering in the reader/viewer the process of imagination and allowing them to develop the own creative thinking on the subject (Lagrasta et al., 2022).

The focus of this work is on how analyst and patient can be 'combined' to create interpsychic transitions. In the perspective of Complex Systems, it could be considered as a 'systemic coupling' that favours an evolution towards an increase in the possibilities of choice (Lenti, 2021, 32). The proposal of this kind of transdisciplinary correlation can contribute to the reflection on the comparison and convergences between Psychoanalysis and Complex Systems.

I would like to point out that a solid psychoanalytic identity and culture allow for fruitful exploration of fields of research not tied to traditional psychoanalytic references. It is hoped that theoretical–clinical research in Psychoanalysis can continue to be enriched by further and more systematic contributions in relation to Complex Systems theory. Psychoanalysis needs to be interfaced with observations from infant research, neuroscience, cognitive and systemic sciences, and artificial intelligence studies. Such developments will allow for a more effective and up-to-date understanding, not only of the interactions between body and mind within complex dynamic interactions, but also of psychopathologies seen in relation to dysfunctionality between different subsystems (Falci, 2021).

References

Battaglia, D., et al. (2022). Questioni di tecnica. In E. Marchiori, L. Masina, & G. Vandi (Eds.), *Elogio della gattaiola. Esplorazioni intorno all'interpsichico* (pp. 15–25).

Bolognini, S. (2008). *Passaggi segreti. Teoria e tecnica della relazione interpsichica.*

Bolognini, S. (2014). Interpsychique, intersubjectif, interpersonnel: états et passages. *Revue Française de Psychosomatique, 45*, 143–161.

Bolognini, S. (2015). Between sameness and otherness. The analyst's words in interpsychic dialogue. In A. Joyce (Ed.), *D.W. Winnicott and the History of* (pp. 17–30).

Bolognini, S. (2016a). The interpsychic dimension in the psychoanalytic interpretation. *Psychoanalytic Inquiry, 36*, 102–111.

Bolognini, S. (2016b). Elementi di tecnica tra Sé e non-Sé. *Rivista di Psicoanalisi, 63*, 5–26.

Bolognini, S. (2019). *Flussi vitali tra Sé e Non-Sé. L'interpsichico.*

Bolognini, S. (2022). Il concetto di interpsichico. In E. Marchiori, L. Masina, & G. Vandi (Eds.), *Elogio della gattaiola. Esplorazioni intorno all'interpsichico* (pp. 1–15).

Bucci, W. (2021). *Emotional communication and therapeitic change. Understanding psychoterapy through multiple code theory.*

De Robertis, D. (2005). Le logiche dei Sistemi Complessi: un potenziale per la teoria e la clinica psicoanalitica. *Ricerca Psicoanalitica, XVI*(3), 319–330.

Falci, A. (2021). Il Nobel 2021. Parisi e i sistemi complessi. https://www.spiweb.it/la-ricerca/ricerca/ricerca-empirica/il-nobel-2021-parisi-e-i-sistemi-complessi-amedeo-falci/, last consultation date 27/5/2023.

Freud, S. (1905). Tre saggi sulla teoria sessuale. *O.S.F.*, 4. Boringhieri, Torino.

Hunter, A. (1994). *Pyschoanalists talk.* Guilford Press.

Lagrasta, D., Marchiori, E., Rosso, C., & Vandi, G. (2022). Scene dell'interpsichico e dei suoi equivalenti intercorporei. In E. Marchiori, L. Masina, & G. Vandi (Eds.), *Elogio della gattaiola. Esplorazioni intorno all'interpsichico* (pp. 27–37).

Lenti, G. (2014). *Psicoanalisi e teoria della complessità nella scienza contemporanea.*

Lenti, G. (2021). *Complessità e campo psicoanalitico. Scritti.*

Marchiori, E., Masina, L., & Vandi, G. (Eds.), (2022a). *Elogio della gattaiola. Esplorazioni intorno all'interpsichico.*

Marchiori, E., Masina, L., & Vandi, G. (2022b). Introduzione. In E. Marchiori, L. Masina, & G. Vandi (Eds.), *Elogio della gattaiola. Esplorazioni intorno all'interpsichico* (pp. V–VII).

Quinodoz, J. (2001). *Les rêves qui tournent une page.* Presses Universitaires de France.https://doi.org/10.3917/puf.quin.2001.01

Racalbuto, A. (1994). *Tra il dire e il fare.*

Complexity of the Academic System: Retention and Dropout

Maria Lidia Mascia, Federica Siddu, and Maria Pietronilla Penna

Abstract The Italian education system is made up of subsystems that should interact with each other to ensure the student's learning and success. Sometimes something does not work, and once compulsory schooling is over, a phenomenon often arises, especially widespread in Italy: school dispersion or dropout. There is evidence of a blockage in the transitions between the various school orders. This difficulty becomes more evident in the period of transition from high school to university, the student finds himself in a system with different characteristics from the previous systems, so that he finds the academic system and manifests discomfort in learning and studying, even to the point of abandoning the path, without obtaining a degree. This system required the student to have a high level of self-regulation. The online laboratory that we have created helps the student acquire greater self-regulation during the first year of university. In this paper, we show a text analysis of the student's thoughts on what he/she experienced.

1 Introduction

The Italian education system is in turn characterised by several subsystems, often very different from each other. In particular, the gap between the high school system and the academic system appears clear in the literature. There are many factors that differentiate the two systems to lead to a problematic transition, which if not well managed can lead the student to drop out of studies before obtaining the degree (Bussu et al., 2020). Dropping out refers to a student who abandons university studies before completing the study program and obtaining the degree (Tinto, 1975). Student

M. L. Mascia (✉) · F. Siddu · M. P. Penna
Department of Pedagogy, Psychology, Philosophy, University of Cagliari, Cagliari, Italy
e-mail: marialidia.mascia@unica.it

F. Siddu
e-mail: federica.siddu0@gmail.com

M. P. Penna
e-mail: penna@unica.it

G. Minati and M. Pietronilla Penna (eds.), *Multiple Systems*, Contributions to Management Science, https://doi.org/10.1007/978-3-031-44685-6_20

dropout is generated by a long decision-making process and the complex interaction between several factors (Behr et al., 2020). The intention to drop out of university is the result of a complex, dynamic, cumulative and multifactorial process. This is also emphasised in a literature review (De Witte et al., 2013) which highlights the complex interaction between individual, organisational and social factors that come into play in this process (Cattelino et al., 2019; Morelli et al., 2021). Studies on this field underline the lack of support for student persistence in the university system associated with a small number of studies investigating the set of variables affecting student success. In this scenario, therefore, the need emerges for an analysis and enhancement of the variables that in the literature are considered fundamental for student persistence and success and that can prevent dropout.

2 Dropout and Self-regulation

Many studies suggest that compared to previous school subsystems, the university system requires more initiative and self-regulation from students, when not possessed sufficiently, this can be a cause of dropout (Suhlmann et al., 2018). Many studies have confirmed the positive influence of a good level of self-regulation on academic success (Wang et al., 2022; Schunk et al., 2012). The literature shows that highly self-regulated students use most appropriate strategies to know what they want to learn, plan and control their learning progress, control their results and, if necessary, adjust or modify their objectives on the light of what they have tried (Schunk et al., 2012; Schumacher & Ifenthaler, 2021). University students are expected to develop critical thinking abilities and reinforce their learning and cognitive skills through behaviours such as resilience, self-regulation, self-discipline and self-motivation in order to face the difficulties they experience in their academic path (Holzer et al., 2021; Ramli et al., 2018).

3 Online Laboratories to Experience Self-regulation

In the literature, positive results in terms of self-regulation have been found with interventions involving problem solving and reflection situations (Hussin et al., 2019) and through the proposal of study and in-depth study activities in open and flexible environments that allow students to be at the centre of the learning process (learner-centred model), as well as to experience the common construction of knowledge, realising collaborative activities in groups (learner-centred team model) (Verstege et al., 2019). The training offers delivered through platforms provide students with optimal activities for their self-regulatory process, combining autonomy in study activities with a constant opportunity for comparison and reflection. Such initiatives imply on the one hand as a prerequisite for good self-regulatory competence, since the student must deal with situations characterised by a high degree of autonomy in learning, by a high degree of autonomy in the management of activities, but on the other hand, the ability to study in a self-regulated manner is frequently stimulated, activated, challenged and supported by participation in activities of a collaborative activities (Núñez Pérez et al., 2011). Also, Azevedo and Hadwin (2005) argue that self-regulation can be supported through the creation and structuring of favourable environments that allow control over the essential dimensions of learning and provide opportunities for reflection and revision. This is since the online environment can provide conceptual, metacognitive, procedural and strategic support.

These environments prove effective both for the maintenance of self-regulatory skills in higher schools and for the strengthening and development of self-regulatory skills during the university course (Urbina et al., 2021).

4 Online Experiences: Qualitative Results

Based on these assumptions, the objective was to carry out an analysis of the participation in the activities proposed in the forums (learning content forums activated by the tutor, spontaneous request forums, support forums and reflection forums) and, through qualitative data processing, to understand the steps progressive steps in the strengthening of their self-regulatory abilities and the requests or problems that emerged most frequently, relating not only to teaching but, in general, to the university organisational system. An attempt was made to understand whether the online tool made available was able to make a significant contribution to the student's self-regulatory and metacognitive abilities, at the same time asking whether the workshop really played a supporting and monitoring role during the student's first year of life. Within the Moodle space, the students were able to reorganise the shared knowledge of the community and propose and/or debate on food for thought related to each of the topics examined. The textual analysis was carried out using the T-Lab software (Lancia, 2004). Textual analysis proves useful to go deeper into the study topic by bringing out sometimes unexpected semantic and thematic dimensions underlying the textual data themselves. The sample was composed as follows: 47 participants for group A; 85 participants for group B; and 96 participants for group C.

An initial analysis was aimed at analysing the text corpus used by the participants during the three editions of the workshop. We were concerned to observe whether the increase in participation also had the effect of increasing the number of interactions and of the text used. During the course, all students were asked for comments aimed at monitoring the study activities in relation to initial expectations. Based on what has been identified in the literature (Araka et al., 2020), the text was divided based on four macro-categories: expectations, organization, self-regulation and content.

For each group, vocabulary emerged in relation to the forums on self-regulation, i.e., where the student was asked to reflect on the relationship between their own study and the workshop. They show the perception of their self-regulation improvement.

5 Conclusion

The data from the evaluations carried out on the online laboratory experience highlighted the perceived benefits of the students involved in the implementation of such activities and is useful in promoting a more self-regulated approach to study. The analysis of the vocabulary spontaneously used highlighted the expectations as well as the perceived benefits following the realisation of the experience. The increase in participation over the subsequent years, basically based on the dissemination by 'word of mouth' of the existence of this opportunity, is in line with this trend.

Of course, these qualitative data do not claim to be exhaustive or explanatory regarding the complex and multifactorial processes involved in the activities of online support activities alongside face-to-face teaching.

It would have been useful, for example, to compare these quantitative data with those relating to a group of students who did not attend the workshop, but this becomes very difficult to achieve.

It would also have been desirable to place the dimensions identified and the processes investigated within the framework of a broader model, which takes into consideration a multiplicity of aspects linked to both face-to-face and online teaching at the basis of student persistence in academia. Another interesting analysis could be the one made, according to recent studies, using learning analytics (Ifenthaler et al., 2019). Future developments include the need to support the student especially in the transition phase, to continue to question the complexity of the school system, to identify predictors of retention and success and to act on these predictors with both in-presence and online support.

References

Araka, E., Maina, E., Gitonga, R., & Oboko, R. (2020). Research trends in measurement and intervention tools for self-regulated learning for e-learning environments—systematic review (2008–2018). *Research and Practice in Technology Enhanced Learning, 15*, 1–21.

Azevedo, R., & Hadwin, A.F. (2005). Introduction to special issue: Scaffolding self-regulated learning and metacognition: Implications for the design of computer-based scaffolds. *Instructional Science, 33*.

Behr, A., Giese, M., Teguim Kamdjou, H. D., & Theune, K. (2020). Dropping out of university: A literature review. *Review of Education, 8*, 614–652.

Bussu, A., Detotto, C., & Serra, L. (2020). Indicators to prevent university drop-out and delayed graduation: An Italian case. *Journal of Applied Research in Higher Education, 12*(2), 230–249.

Cattelino, E., Morelli, M., Baiocco, R., & Chirumbolo, A. (2019). From external regulation to school achievement: The mediation of self-efficacy at school. *Journal of Applied Developmental Psychology, 60*, 127–133.

De Witte, K., Cabus, S., Thyssen, G., Groot, W., & van Den Brink, H. M. (2013). A critical review of the literature on school dropout. *Educational Research Review, 10*, 13–28.

Holzer, J., Lüftenegger, M., Korlat, S., Pelikan, E., Salmela-Aro, K., Spiel, C., & Schober, B. (2021). Higher education in times of COVID-19: University students' basic need satisfaction, self-regulated learning, and well-being. *AERA Open, 7*, 1–13.

Hussin, W. N. T. W., Harun, J., & Shukor, N. A. (2019). A review on the classification of students' interaction in online social collaborative problem-based learning environment: How can we enhance the students' online interaction. *Universal Journal of Educational Research, 7*(9A), 125–134.

Ifenthaler, D., Mah, D. K., & Yau, J. Y. K. (Eds.). (2019). *Utilizing learning analytics to support study success.* Springer.

Lancia, F. (2004). *Strumenti per l'analisi dei testi. Introduzione all'uso di T-LAB.* FrancoAngeli.

Morelli, M., Chirumbolo, A., Baiocco, R., & Cattelino, E. (2021). Academic failure: Individual, organizational, and social factors. Psicología Educativa. *Revista de los Psicólogos de la Educación, 27*(2), 167–175.

Núñez Pérez, J. C., Cerezo Menéndez, R., Bernardo Gutiérrez, A. B., Rosário, P., Valle Arias, A., Fernández Alba, M. E., & Suárez Fernández, N. (2011). Implementation of training programs in self-regulated learning strategies in Moodle format: Results of an experience in higher education. *Psicothema.*

Ramli, N. H. H., Alavi, M., Mehrinezhad, S. A., & Ahmadi, A. (2018). Academic stress and self-regulation among university students in Malaysia: Mediator role of mindfulness. *Behavioral Science, 8*, 12.

Schumacher, C., & Ifenthaler, D. (2021). Investigating prompts for supporting students' self-regulation—A remaining challenge for learning analytics approaches? *Internet High Education, 49*, 100791.

Schunk, D. H., & Zimmerman, B. J. (2012). Motivation and self-regulated learning: Theory, research, and applications. In *Motivation and self-regulated learning: Theory, research, and applications* (pp. 1–417). Lawrence Erlbaum Associates.

Suhlmann, M., Sassenberg, K., Nagengast, B., & Trautwein, U. (2018). Belonging mediates effects of student-university fit on well-being, motivation, and dropout intention. *Social Psychology, 49*, 16–28.

Tinto, V. (1975). Dropout from higher education: A theoretical synthesis of recent research. *Review of Educational Research, 45*(1), 89–125.

Urbina, S., Villatoro, S., & Salinas, J. (2021). Self-regulated learning and technology-enhanced learning environments in higher education: A scoping review. *Sustainability, 13*(13), 7281.

Verstege, S., Pijeira-Díaz, H. J., Noroozi, O., Biemans, H., & Diederen, J. (2019). Relations between students' perceived levels of self-regulation and their corresponding learning behavior and outcomes in a virtual experiment environment. *Computers in Human Behavior, 100*, 325–334.

Wang, H., Yang, J., & Li, P. (2022). How and when goal-oriented self-regulation improves college students' well-being: A weekly diary study. *Current Psychology, 41*, 7532–7543.